Bioreactor Design and Product Yield

BOOKS IN THE BIOTOL SERIES

The Molecular Fabric of Cells
Infrastructure and Activities of Cells

Techniques used in Bioproduct Analysis
Analysis of Amino Acids, Proteins and Nucleic Acids
Analysis of Carbohydrates and Lipids

Principles of Cell Energetics
Energy Sources for Cells
Biosynthesis and the Integration of Cell Metabolism

Genome Management in Prokaryotes
Genome Management in Eukaryotes

Crop Physiology
Crop Productivity

Functional Physiology
Cellular Interactions and Immunobiology
Defence Mechanisms

Bioprocess Technology: Modelling and Transport Phenomena
Operational Modes of Bioreactors

In vitro Cultivation of Micro-organisms
In vitro Cultivation of Plant Cells
In vitro Cultivation of Animal Cells

Bioreactor Design and Product Yield
Product Recovery in Bioprocess Technology

Techniques for Engineering Genes
Strategies for Engineering Organisms

Principles of Enzymology for Technological Applications
Technological Applications of Biocatalysts
Technological Applications of Immunochemicals

Biotechnological Innovations in Health Care

Biotechnological Innovations in Crop Improvement
Biotechnological Innovations in Animal Productivity

Biotechnological Innovations in Energy and Environmental Management

Biotechnological Innovations in Chemical Synthesis

Biotechnological Innovations in Food Processing

Biotechnology Source Book: Safety, Good Practice and Regulatory Affairs

BIOTECHNOLOGY BY OPEN LEARNING

Bioreactor Design and Product Yield

PUBLISHED ON BEHALF OF:

Open universiteit and **Thames Polytechnic**

Valkenburgerweg 167
6401 DL Heerlen
Nederland

Avery Hill Road
Eltham, London SE9 2HB
United Kingdom

Butterworth-Heinemann Ltd
Linacre House, Jordan Hill, Oxford OX2 8DP

 PART OF REED INTERNATIONAL BOOKS

OXFORD LONDON BOSTON
MUNICH NEW DELHI SINGAPORE SYDNEY
TOKYO TORONTO WELLINGTON

First published 1992

British Library Cataloguing in Publication Data
A catalogue record for this book is
available from the British Library

Library of Congress Cataloguing in Publication Data
A catalogue record for this book is
available from the Library of Congress

ISBN 0 7506 1509 5

Printed and bound in Great Britain by
Thomson Litho Ltd, East Kilbride, Scotland

The Biotol Project

The BIOTOL team

OPEN UNIVERSITEIT, THE NETHERLANDS
Prof M. C. E. van Dam-Mieras
Prof W. H. de Jeu
Prof J. de Vries

THAMES POLYTECHNIC, UK
Prof B. R. Currell
Dr J. W. James
Dr C. K. Leach
Mr R. A. Patmore

This series of books has been developed through a collaboration between the Open universiteit of the Netherlands and Thames Polytechnic to provide a whole library of advanced level flexible learning materials including books, computer and video programmes. The series will be of particular value to those working in the chemical, pharmaceutical, health care, food and drinks, agriculture, and environmental, manufacturing and service industries. These industries will be increasingly faced with training problems as the use of biologically based techniques replaces or enhances chemical ones or indeed allows the development of products previously impossible.

The BIOTOL books may be studied privately, but specifically they provide a cost-effective major resource for in-house company training and are the basis for a wider range of courses (open, distance or traditional) from universities which, with practical and tutorial support, lead to recognised qualifications. There is a developing network of institutions throughout Europe to offer tutorial and practical support and courses based on BIOTOL both for those newly entering the field of biotechnology and for graduates looking for more advanced training. BIOTOL is for any one wishing to know about and use the principles and techniques of modern biotechnology whether they are technicians needing further education, new graduates wishing to extend their knowledge, mature staff faced with changing work or a new career, managers unfamiliar with the new technology or those returning to work after a career break.

Our learning texts, written in an informal and friendly style, embody the best characteristics of both open and distance learning to provide a flexible resource for individuals, training organisations, polytechnics and universities, and professional bodies. The content of each book has been carefully worked out between teachers and industry to lead students through a programme of work so that they may achieve clearly stated learning objectives. There are activities and exercises throughout the books, and self assessment questions that allow students to check their own progress and receive any necessary remedial help.

The books, within the series, are modular allowing students to select their own entry point depending on their knowledge and previous experience. These texts therefore remove the necessity for students to attend institution based lectures at specific times and places, bringing a new freedom to study their chosen subject at the time they need and a pace and place to suit them. This same freedom is highly beneficial to industry since staff can receive training without spending significant periods away from the workplace attending lectures and courses, and without altering work patterns.

Contributors

AUTHORS

Dr G. Mijnbeek, Bird Engineering bv, Schiedam, The Netherlands

Dr Ir N. M. G. Oosterhuis, Bird Engineering bv, Schiedam, The Netherlands

Dr M. A. Siebel, IHE Delft, Delft, The Netherlands

Dr M. C. E. van Dam-Mieras, Open universiteit, Heerlen, The Netherlands

Dr Ir R. T. J. M. van der Heijden, University of Technology Delft, Delft, The Netherlands

TECHNOLOGY AND EDITORIAL ADVISORS

Dr R. O. Jenkins, Leicester Polytechnic, Leicester, UK

Dr C. K. Leach, Leicester Polytechnic, Leicester, UK

Dr G. Mijnbeek, Bird Engineering BV, Schiedam, The Netherlands

SENIOR TECHNOLOGY ADVISOR

Professor Ir K. Ch. A. M. Luyben, University of Technology Delft, Delft, The Netherlands

SCIENTIFIC AND COURSE ADVISORS

Professor M. C. E. van Dam-Mieras, Open universiteit, Heerlen, The Netherlands

Dr C. K. Leach, Leicester Polytechnic, Leicester, UK

ACKNOWLEDGEMENTS

Grateful thanks are extended, not only to the authors, editors and course advisors, but to all those who have contributed to the development and production of this book. They include Miss K. Brown, Dr G. M. Hall, Dr M. de Kok, Miss J. Skelton and Professor R. Spier.

The development of this BIOTOL text has been funded by **COMETT, The European Community Action Programme for Education and Training for Technology**. Additional support was received from the Open universiteit of The Netherlands and by Thames Polytechnic.

Project Manager: Dr J. W. James

Contents

How to use an open learning text

An open learning text presents to you a very carefully thought out programme of study to achieve stated learning objectives, just as a lecturer does. Rather than just listening to a lecture once, and trying to make notes at the same time, you can with a BIOTOL text study it at your own pace, go back over bits you are unsure about and study wherever you choose. Of great importance are the self assessment questions (SAQs) which challenge your understanding and progress and the responses which provide some help if you have had difficulty. These SAQs are carefully thought out to check that you are indeed achieving the set objectives and therefore are a very important part of your study. Every so often in the text you will find the symbol Π, our open door to learning, which indicates an activity for you to do. You will probably find that this participation is a great help to learning so it is important not to skip it.

Whilst you can, as a open learner, study where and when you want, do try to find a place where you can work without disturbance. Most students aim to study a certain number of hours each day or each weekend. If you decide to study for several hours at once, take short breaks of five to ten minutes regularly as it helps to maintain a higher level of overall concentration.

Before you begin a detailed reading of the text, familiarise yourself with the general layout of the material. Have a look at the contents of the various chapters and flip through the pages to get a general impression of the way the subject is dealt with. Forget the old taboo of not writing in books. There is room for your comments, notes and answers; use it and make the book your own personal study record for future revision and reference.

At intervals you will find a summary and list of objectives. The summary will emphasise the important points covered by the material that you have read and the objectives will give you a check list of the things you should then be able to achieve. There are notes in the left hand margin, to help orientate you and emphasise new and important messages.

BIOTOL will be used by universities, polytechnics and colleges as well as industrial training organisations and professional bodies. The texts will form a basis for flexible courses of all types leading to certificates, diplomas and degrees often through credit accumulation and transfer arrangements. In future there will be additional resources available including videos and computer based training programmes.

Preface

The popular misconceptions that biotechnology is synonymous with genetic manipulation and cell cloning belittles the importance of process technology to the successful development of biotechnology. Biotechnology is much more than a sub-discipline of biology. Although it is true that new knowledge emerging from the relevant biological sciences, has provided a fresh impetus to biotechnological exploitation over the past few decades, the key to successful exploitation lies in developments in the associated process technology. The integration of process engineering strategies with the possibilities made available by scientific research is essential if the desired goals of improved health care, food manufacture, agriculture and environmental protection are to be realised. Recent years have seen much progress made in the relevant engineering practices and we are rapidly moving away from using the empirical techniques of the past to a rationally developed technology much more fitted to exploit the biological possibilities.

This text is one of four BIOTOL texts designed especially to develop the knowledge and understanding of the principles and practices of contemporary bioprocess technology. These four texts are:

- Bioprocess Technology: Modelling and Transport Phenomena
- Operational Modes of Bioreactors
- Bioreactor Design and Product Yield
- Product Recovery in Bioprocess Technology

The first two of these texts deal with the fundamental principles which underpin bioreactor performance. This text extends this development to examine the issues of scale up and product yield and also includes a short introduction to process control. The final text in the series explains the processes and strategies that can be used for the recovery of products from bioreactors.

In this text, the authors have provided a brief orientating chapter which explain what pre-knowledge is assumed of the reader and to more fully explain the layout of the text. This is followed by a section dealing with the strategies that can be used to scale up laboratory scale processes to production scale. This section mainly examines the selection of scale up criteria and the advantages and limitations inherent in using such criteria. It also introduces the newly developing technique referred to as scale down. This section leads on to consider the performance of bioreactors of different design and configuration. The main focus is on the yields of biomass and other bioproducts and their relationships with process parameters and variables. The final part of the text introduces aspects of process control. The main focus here is on simple linear control systems.

The contributors have made an excellent job of integrating logically developed technical themes with sound educational practice. This text is a learning resource of distinction. It has made accessible to the reader, much of the current thinking in bioprocess technology, a reflection of the status of authors and advisors and the care they took in selecting and developing material. Our thanks to all who have contributed. It is up to the reader to take full advantage of the opportunities this text offers. We encourage you to attempt the in text activities which have been designed to facilitate your learning and to use the self assessed questions to check your progress. In this way you will maximise the benefit you may gain from the text.

Scientific and Course Advisors: Prof M. C. E. van Dam-Mieras

Dr C. K. Leach

An introduction to bioreactor design and product yield

An introduction to bioreactor design and product yield

1.1 Sectors in the biotechnology industry

Increased knowledge of how biological systems function has undoubtedly increased mankind's ability to manipulate these systems. Key to this manipulation has been the development of the techniques which enable us to transfer genes from one organism to another and to modify the expression of genes within organisms. These techniques, commonly referred to as gene manipulation or genetic engineering are not simply confined to re-arranging existing, naturally occurring genes. Increasingly, through the use of molecular biological approaches, genes themselves are being modified to make more desirable products. For example by making the product made by the gene easier to purify or more stable and, therefore, longer-lived in use.

importance of bioprocess technology

The advances in biotechnology are not, however, solely a consequence of increase in biological knowledge and expertise. Of equal importance has been the progress made in process engineering. Industrial biotechnology, in essence, makes use of biological systems to catalyse chemical changes. The types of chemical changes that are of interest are very diverse. In some cases the value to mankind is to remove undesired components from a system. Waste water treatment is a good example of this. In such a process we seek to remove organic (and other) materials in the water to generate a non-toxic, usable product (clean water). In other cases we are interested in the chemical products made by the biological system. These may be bulk products like ethanol, citric acid or biomass itself or they may be fine chemicals including pharmaceutical compounds such as antibiotics and hormones. The diversity of biotechnological products and processes is described in Table 1.1.

diversity of products

Π Examine Table 1.1 carefully and estimate (or guess) for each product, the volume produced per year.

Industrial sector		Examples of products and processes
Pharmaceuticals		vaccines antibiotics diagnostics steroids alkaloids protein hormones and blood factors
Food		dairy meat and fish products beverages (alcohol, coffee, tea, etc) bread, bakers yeast food additives (antioxidants, colours, flavours) food supplements (amino acids, vitamins) starch products glucose and high fructose syrups enzymes novel foods, fungi
Agriculture		plant cell and tissue culture for quality assured stock genetically improved varieties biopesticides ensilaging and composting animal feedstuff manufacture vaccines
Chemicals	(bulk)	ethanol, butanol organic acids metal extraction
	(fine)	enzymes polymers (gums,agars) perfumes
Energy		ethanol methane biomass
Service		efficient treatment water re-claimation oil recovery analysis

Table 1.1 Industrial sectors and biotechnology products and processes.

We would anticipate that you would not have very accurate figures for each but you probably have placed each in the categories of fairly small, medium and large. In practice high priced, low volume products such as some of the specialised pharmaceuticals are produced in the range of $0.1\text{-}10^2$ kg y^{-1}. In the service sector, volumes of effluent treated is in the range of $10^6\text{-}10^{10}$ m^3 y^{-1}. There is, in fact, an enormous range in the scales of biotechnological processes.

We can identify three quite distinct scales of operation. Table 1.2 describes the broad characteristics of these three sectors.

Although we have divided these into three groups, you should realise that there are some areas of overlap. For example the use of recombinant DNA (genetically engineered) organisms are not included under sector III, but there is no inherent reason why they should not be used on this scale.

Characteristic	Sector I	Sector II	Sector III
Volumes	$0.1\text{-}10^2$ kg y^{-1}	$10^3\text{-}10^5$ kg y^{-1}	$10^6\text{-}10^9$ kg y^{-1}
Organism	rDNA	partly rDNA	natural producers
Product purity	very high	high/very high	relatively low
Recovery yield	subordinate importance	of minor importance	highly important
Cost price	fraction	20-50% determined by raw materials	50-90% determined by raw materials

Table 1.2 Characteristics of the three sectors of biotechnology based on the volume of annual output (rDNA = recombinant DNA, ie genetically engineered strains).

Π Examine Tables 1.1 and 1.2 carefully and mark on the tables those products and processes which involve the application of process technology be it for primary production, downstream processing or product transport and packaging.

You may not have been familiar with some of the products listed in Table 1.1. Nevertheless, a moments thought will have enabled you to realise that virtually all biological products will require the application of process technology. In most cases, the requirements of the market, the nature of the product and the need to retain commercial viability, results in the application of process technology to processes carried out on a large scale. Examination of Table 1.1 could not fail to have impressed you not only by the great diversity and value of bioproducts, but also by the importance of process technology in enabling us to produce these bioproducts.

Bioprocess technology essentially turns biological possibilities into practical propositions.

1.2 The components of bioprocess technology and the BIOTOL texts

Process biotechnology has its origins in chemical engineering. In both areas we are predominantly concerned with transforming one set of chemicals into another. The fundamental difference is that when we are using biological agents to carry out the catalysis, we can bring about more complex chemical changes with greater precision and yield. This advantage is, however, partially counterbalanced by the very specific demands that biological systems make on the environment in which they operate.

three main stages used in bioprocess technology

For example they will only operate over a limited temperature and pH range and almost always have to be in an aqueous milieu. Thus in biotechnological processes we might anticipate a narrow band of reactor types, all operating at more-or-less the same temperature and with an aqueous milieu. Nevertheless we can, in common with chemical engineering, identify three main stages in bioprocess. These are:

- upstream processes (eg preparation of media, organisms, etc);
- reactor (stage in which the bioconversion processes take place);
- downstream processes (stage in which the desired product is recovered from the reactor milieu).

The balance between these three stages of course differs for different products. Thus with products that need to be used in a highly purified form (eg pharmaceuticals) downstream processing can be quite extensive. In others, such as waste water treatment, downstream processing may be quite simple.

Irrespective of the balance of these three stages, or on the exact scale of operation, we can recognise some principles which underpin each stage of the process. Predominant amongst these are the principles involved in transfer (transport) processes. It is self-evident that we have to get nutrients (substrates) to the biocatalysts, remove products from the milieu and so on. Thus underpinning all stages in bioprocesses is the need to understand transport (including both mass and energy) phenomena and to be able to write mass and energy balances. This underpinning knowledge is provided by the BIOTOL text 'Bioprocess Technology: Modelling and Transport Phenomena'.

In developing schemes in bioprocess technology, it is also helpful to be able to model the process under development. The modelling of microbiological processes and the kinetics of conversion rates are also described in the BIOTOL text referred to above.

A vital element in bioprocesses is, of course, the bioreactor, the place where the bioconversion takes place. Bioreactors are dealt with in two BIOTOL texts. The first one, entitled 'Operational Modes of Bioreactors' builds upon the fundamental principles of transport and modelling covered in 'Bioprocess Technology Modelling and Transport Phenomena' and applies these principles to the commonly encountered bioreactor types. Particular emphasis is placed on the transfer processes (both at the micro and macro levels) which occur in these reactors. Oxygen availability is a particular concern in aerobic systems. Thus this second BIOTOL process technology text begins by examining some basic parameters used in the description of oxygen transport. This knowledge is then applied to consideration of the properties of each reactor type. Likewise the discussion of immobilised systems centres on consideration of the principles of mass transfer within these systems. The text finishes by discussing the strategies for sterilisation in virtually all bioreactor operations.

The text in hand is the second BIOTOL text devoted to bioreactors. It utilises basic knowledge of transport phenomena, modelling and bioreactor types to develop strategies for evaluating important process parameters. In practical terms, we are either interested in the rate of substrate removal (eg in waste water treatment) or biomass and/or product yield. The emphasis of this text is therefore on these aspects of bioreactor performance. The text begins by developing strategies for transforming and interpreting data from small scale operations to large production processes. It then develops into more specific discussions of the relationships between substrate consumption, biomass production and product formation together with other process parameters. The final phase of the text examines the principles of process control and introduces the reader to a variety of process control strategies and devices. Although many examples are included, the approach used is a generic one in which the relationships between measurable parameters and reactor performance are established. This approach provides the reader with knowledge that is applicable in a wide range of circumstances and develops an analytical awareness of the issues involved in bioreactor design and performance.

A further BIOTOL text ('Product Recovery in Bioprocess Technology') deals with downstream processing. This text covers many issues including the characteristics of the fermentation broths, the release of intracellular components, solid-liquid separation, concentration and purification.

Thus within the BIOTOL series we identify four supporting texts which will provide an understanding of bioprocess technology. We can represent these by the following scheme:

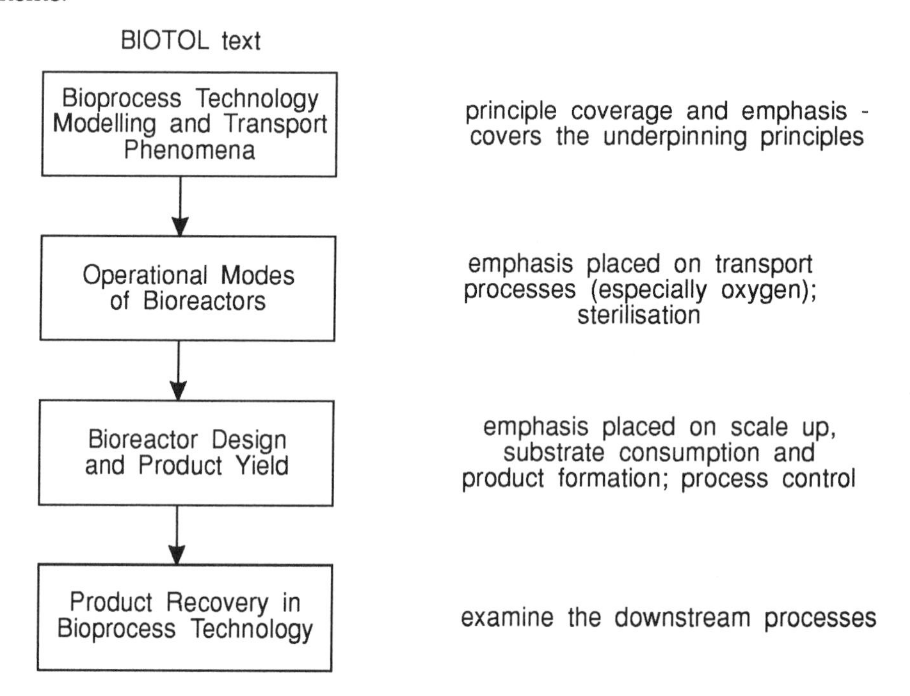

1.3 Assumed knowledge

From the description given in Section 1.2, it should be clear that this text has been written on the assumption that the reader has knowledge of the underpinning principles of mass and energy transfer (transport), modelling and basic reactor types. However, the authors have provided many helpful ' reminders' and help lines. Thus although your knowledge of transfer process may be limited or a little rusty, the open, friendly style of this text will make it accessible. Every effort has been made to foresee likely difficulties and to build in remedial help within the text. Nevertheless, if you only have scant knowledge of mass and energy transfer processes and have no knowledge or experience of dimensional analysis, we would recommend you tackle the BIOTOL text 'Bioprocess Technology: Modelling and Transport Phenomena' before embarking on this text.

Strategies for scale up

Strategies for scale up

2.1 General introduction

In general, scale up of a fermentation process occurs when a new or an improved production process has to be transferred from laboratory scale (0.2-5.0 l) to production scale (1-200 m^3). On the production scale either an existing bioreactor or a specially designed bioreactor can be used. The designing of a new bioreactor is not an easy task. A large scale bioreactor is not inexpensive, and in setting up a process system, it is essential to consider future needs and to estimate the degree of scaling up that will be required. Furthermore, a fermentation process developed initially in the laboratory may not easily be scaled up to full production because it is difficult to assess certain factors affecting the scale up process. As a result, many large scale fermentation processes give a lower yield than is expected from laboratory scale experiments, or the scaled up reactor is over-designed. In both cases this means higher production costs than for a good optimised and scaled process. Therefore, a laboratory bioreactor is usually scaled up first to a pilot process plant, so that various factors involved in full production can be considered and evaluated, and the necessary adjustments can be made to the process. This procedure clearly shows the existence of uncertainties in scale up.

Some authors claim that the stirred tank reactor is a well proven technology for carrying out biotechnological processes and that improvements in reactor design are only needed for production of bulk chemicals or waste water treatment. It could be argued therefore that further research on reactor design for biotechnological processes is a waste of time and money. However, results from practice show the contrary. New reactor designs, resulting in easy possibilities for scale up are not only needed to improve production of bulk chemicals or processes for waste water treatment, but are especially needed to improve production of complex systems such as cultivation of mammalian cells and genetic engineered micro-organisms. The anticipated future use of mammalian cells underlines this need for special reactor design. For the production of monoclonal antibodies for example, a tenfold increase in the market in the coming few years is expected. So, production at larger scales using such complex biosystems will be needed to achieve economical feasibility.

structure of this chapter

In this chapter we will examine the strategies we can adopt to solve the problems of scale up. The chapter is divided into a number of sections. Before we embark on examining the methods used in scale up, we will briefly examine the main factors that need to be considered in scale up. Using this information, we will then examine the methods employed in examining scale up before exploring what has been called regime analysis. In the next chapter we will examine the so-called scale down approach. Scale down, based on regime analysis, is an important tool to scale up microbial processes. This chapter is a long one so do not attempt to do it all in one go. We advise you to take advantage of the in text activities (Π) to aid your learning and to attempt the self assessed questions (SAQs) to check your progress and understanding.

2.1.1 The important phenomena in bioreactor design

Π Before reading on, make a list of the factors which need to be considered in the design of a bioreactor (use a separate piece of paper and spend 5 minutes on this).

bioreactor design

We would be very surprised if you have not generated a very long list. You probably included such items as: how big does the vessel need to be to produce sufficient product; what materials should the vessel be constructed out of; how are the materials in the vessel to be mixed efficiently but without damage to the cells in the vessel, what are the costs (capital, running, depreciation) of various types of equipment and so on. There are, indeed, a great many issues involved in bioreactor design. Here we will focus on the important phenomena which must underpin our decision-making in terms of reactor design. We identify three different phenomena of great importance. These phenomena are:

- thermodynamics;
- microkinetics;
- transport.

oxygen concentration and kinetic behaviour depend on scale

These three phenomena essentially govern the performance of a bioreactor. Of these phenomena the first and second are independent of scale (if measured properly). Neither a typical thermodynamic property like the solubility of oxygen in a broth nor microkinetic properties (for example growth and product formation by the micro-organism) are dependent on the scale of the bioreactors. However, the actual oxygen concentration and the kinetic behaviour of micro-organisms in a bioreactor are rather dependent on scale. The reason is that oxygen and other nutrients involved in the conversion processes are consumed constantly and have to be supplied by transport processes. These transport processes are influenced by scale up which can result in the existence of gradients at production scale. When the micro-organism travels through the bioreactor it encounters constantly changing concentrations of nutrients, oxygen and pressure. Furthermore, the micro-organisms are always subjected to (turbulent) shear phenomena, which are also a kind of transport processes (momentum transport). Shear phenomena can either damage the micro-organism/cell itself or influence the formation of agglomerates (flocs, pellets) of micro-organisms.

Since transport of oxygen and nutrients to micro-organisms is essential in (industrial) fermentations, scale up problems can always be expected (although more for aerobic processes than for anaerobic ones and more for continuous- and fed-batch processes than for batch processes). Transport phenomena are governed mainly by flow and diffusion which are directly related to shear, mixing, mass transfer (k_La), heat transfer and macrokinetics (a form of apparent kinetics as a result of a combination of microkinetics and diffusions: occurring in immobilised systems, flocs and pellets). Therefore, we can expect these phenomena to be liable to change during scale up. The kinetic behaviour of micro-organisms is dependent upon local environmental conditions, and these, in turn, are determined to a great extent by mixing phenomena. Thus one can also expect that the apparent kinetic behaviour can change with scale up.

We remind you that we can consider transport phenomena on a molecular (micro) scale or on a macroscale. Transport at the macroscale will, of course, influence events at the microscale.

We can conclude that basically scale up problems exist when there is a transport of heat, mass or momentum in a system. This can be illustrated by a simple example. Let us consider a plug-flow reactor with dispersion, and a first order homogeneous reaction without temperature gradients. The balance for mass transfer can be written in the following form:

$$\begin{matrix}\text{(Diffusion)} \\ \text{in} - \text{out}\end{matrix} + \begin{matrix}\text{(Convection)} \\ \text{in} - \text{out}\end{matrix} + \text{(Production)} = \text{(Accumulation)} \qquad \text{(E - 2.1)}$$

or

$$D \cdot \frac{d^2C}{dx^2} - v \cdot \frac{dC}{dx} + kC = \frac{dC}{dt} \qquad \text{(E - 2.2)}$$

where:

C = concentration

D = diffusion coefficient

k = reaction rate constant.

v = velocity

The dispersion term ($D\, d^2C/dx^2$) as well as the flow term ($v\, dC/dx$) are place and thus scale dependent. However with free cells, the kinetic term (kC) is scale independent (microkinetics). In cases where we have agglomerates of cells, kinetics are scale dependent too (macrokinetics). It should not be surprising that the bulk of the discussion in this chapter is therefore focused on the issues of transport since these dominate the strategies used in scale up.

You should however note that the success of an improved or new microbial process depends on the scale up rules. However, in contrast to chemical engineering, few fundamental rules or methods for biochemical engineering are known. In this discipline the trial-and-error method is still used. Furthermore, if a new organism is selected for production, it is important to consider whether adaptation of this organism to large scale conditions is necessary. An example is the pellet formation of fungi in penicillin production. It is clearly important to know if the fungus grows in a pelleted form or in a non-aggregated form if we are to predict transport phenomena within the culture. In the next section the classical methods for scale up, most of which have been derived from chemical engineering, and their possible application to biotechnological engineering will be discussed. This will be followed by sections on regime analysis and the scale down approach. Scale down, based on regime analysis, is an important tool to scale up microbial processes; together with some examples it will be discussed in more detail in the next chapter.

2.2 Introduction to scale up methods

A wide variety of methods have been used in scale up. These include:

- fundamental method;
- semi-fundamental method;
- rules of thumb;
- dimensional analysis;

- regime analysis.

In this section we will briefly describe the fundamental and semi fundamental methods and introduce the 'rule of thumb' approach. This latter approach is so important in practice that we will examine aspects of this approach in some detail. Subsequent sections will deal with the dimensional analysis and regime analysis approach to scale up.

2.2.1 Fundamental method

This method for chemical engineering implies solving all the microbalances for momentum-, mass-, and heat-transfer in the system. A number of complications arise when these balances are used for scale up of a fermentation process. Firstly, the balances, if applied to a stirred vessel, have to contain terms for transport in three directions. Furthermore, the boundary conditions are very complicated. Secondly, the balances are coupled: the solution of the momentum balance gives the flow components that have to be used in the mass and heat balances. Thirdly, the momentum balances are usually set up for a homogenous fluid which is not very realistic for an aerated fermentation broth. The main problem is, however, the impossibility of solving the micromomentum balance. For some scale up problems, particularly when the flow is well defined (or absent), solving the microbalances can, however, be very useful.

2.2.2 Semi-fundamental method

Scale up can be based on simplified flow models in order to avoid the use of too complex momentum balance equations. Most models refer to bulk flow and do not give information about flow near important regions such as stirrer blades, cooling coils, and vessel walls. The flow models used are generally one of the following three types: plug flow, plug flow with dispersion or well-mixed systems. Combinations of these models are used as well. Because these bulk flow models usually result from observations in small scale equipment, scale up is based on extrapolation (in other words: is based on the similarity principle) which makes it rather risky. The concept of similarity is used quite often in scale up literature. Four similarity states are important in chemical engineering: geometrical, mechanical, thermal and chemical, each state is dependent on the previous ones (see also Section 2.5). Consequently, similar bulk flow profiles at laboratory and production scale imply geometric similarity of both scales.

2.2.3 Rules of thumb

criteria used for scale up

In biochemical engineering 'rules of thumb' combined with extrapolation to larger scale (using only the geometric similarity principle) are commonly applied as a scale up procedure. Surveys of the criteria used in scale up by the European fermentation industry indicate a rather small range of criteria that have been used (see Table 2.1).

% of industries	scale-up criterion used
30	power per volume ratio
30	oxygen transfer rate (k_La)
20	impeller tip speed
20	oxygen tension

Table 2.1 Scale-up criteria used by some fermentation industries (derived from Einsele, 1978, Proc Biochem 13, 13-14 and Margarites and Zajic, 1978, Biotechnol Bioeng 20, 939-1001).

Π Examine Table 2.1 and decide what is the common feature of the criteria used in scale up.

You should have come to the conclusion that the criteria referred to in Table 2.1 are all closely related and refer mainly to oxygen transfer (the oxygen tension pO_2 is a function of the mass transfer coefficient k_La which in turn is a function of the ratio of power and volume P/V, providing the stirrer tip speed $v_{tip} > 3$ m s^{-1}. This latter relationship between k_La and the P/V ratio is discussed in detail in the BIOTOL text 'Operational Modes of Bioreactors').

Frequently the choice of a criterion is empirical and not well founded by experimental data and model design.

It can be shown that scale up based on empirical criteria is rather complicated (see Table 2.2). In this table the values in every row give, for a particular scale up criterion, the ratio for prototype and model of the variable mentioned at the top of the table. The calculations have been set up for geometrically similar systems with volume of the model (V_M) =10 l and the volume of the prototype (V_P) = 10 m^3. Thus a linear scale up factor is 10. As one can see, different scale up criteria result in entirely different process conditions at production scale. Thus to give an equal P/V value, we have to increase the power input by 10^3. On the other hand to have an equal value of N (stirred speed), the power input has to be increased by a factor of 10^5.

	ratio of the value at 10 m^3 relative to the value at 10 l				
scale-up criterion	P	P/V	N (or t_m^{-1})	ND	Re
equal P/V	10^3	1	0.22	2.15	21.5
equal N (or t_m)	10^5	10^2	1	10^2	10^2
equal tip speed	10^2	0.1	0.1	1	10
equal Re-number	0.1	10^{-4}	0.01	0.1	1

Table 2.2 Different scale-up criteria and their consequences (P = power, V = volume, N = stirrer speed, D = stirrer diameter, Re = stirrer Reynolds number). See text for details.

Π From Table 2.2, how much does the P/V ratio have to be changed to achieve the same Re-number?

From the table, you should have read that to achieve the same Re number in a 10 m^3 vessel as in a 10 l vessel, we would need to change the P/V ratio by 10^{-4}.

Π How much bigger would the diameter of the stirrer need to be if we wished to use the same stirrer speed?

From the table, to use the same N, the value ND must be increased by 10^2. Thus the diameter would need to be increased by 10^2.

Problems which appear when scale up is based on such criteria can be illustrated from literature data. Although we will give full references to these examples, we do not demand that you go and read the original articles. Reading some of these would however provide you with some experience on how these matters are described and discussed in the literature.

As a first example the protease production by a *Streptomyces* sp. is considered. Takei *et al* (1975) made a study of this production in a 0.03 m^3 and a 0.2 m^3 reactor (Takei, M, Mizusawa, K Yoshida, F, 1975, J. Ferment Technol 53, 151-158). The ability of the cells to produce protease is influenced by the oxygen transfer rate (OTR) as can be seen in Figure 2.1.

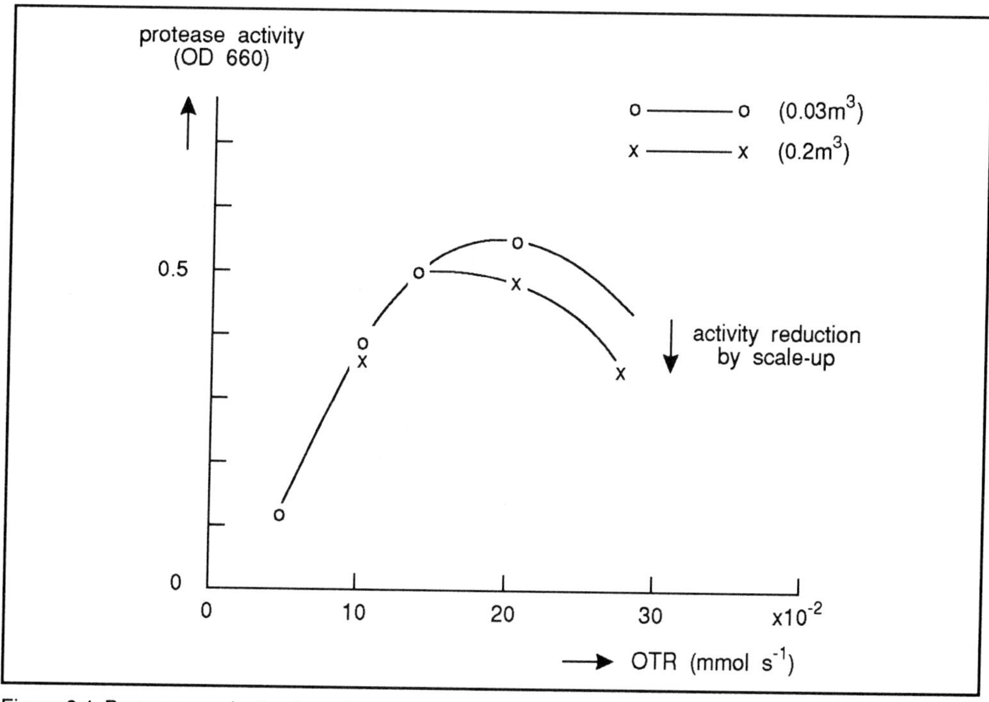

Figure 2.1 Protease production by a *Streptomyces* sp. at different scales (after Takei *et al*, 1975, J Ferm Technol 53, 151-158).

Π Examine Figure 2.1 carefully. Can you conclude from the data presented in this figure, if OTR is the only factor determining protease activity?

You should have concluded that at higher oxygen transfer rates, the protease activity is influenced by scale dependent effects. So, in this case a constant oxygen transfer rate at the different scales is not the only valuable criterion for scale up. Obviously the relationship between OTR and protease activity changes on scale up. Another example is the production of gluco-amylase by an *Endomyces* sp. which is affected by the oxygen transfer rate (Taguchi *et al* 1968 J Ferment Technol 48, 814-826). The data are presented in Table 2.3 and Figure 2.2.

Characteristic		Scale of operation 0.06 m^3	3 m^3	30 m^3	
aeration	(vvm)	1	0.5	0.33	
agitation	(rpm)	400	180	150	
k_La	(min^{-1})	2.28	1.73	1.14	1.58*)
relative activity of gluco-amaylase	(%)	95	100	61	91

Table 2.3 Gluco-amylase production by an *Endomyces* sp. (After Taguchi *et al*, 1968 J Ferm Technol 48, 814-826). The values shown were determined with a straightforward geometric scale-up. Note how when the vessel gets bigger, the k_La value falls, due to the reduced aeration and agitation rates. The value marked * is for a non-geometric scale-up. In this case instead of using an impeller of diameter 0.79m (which it needed to be if the scale-up was to remain geometrically consistent), the impeller was increased to a diameter of 0.94m (see text for details).

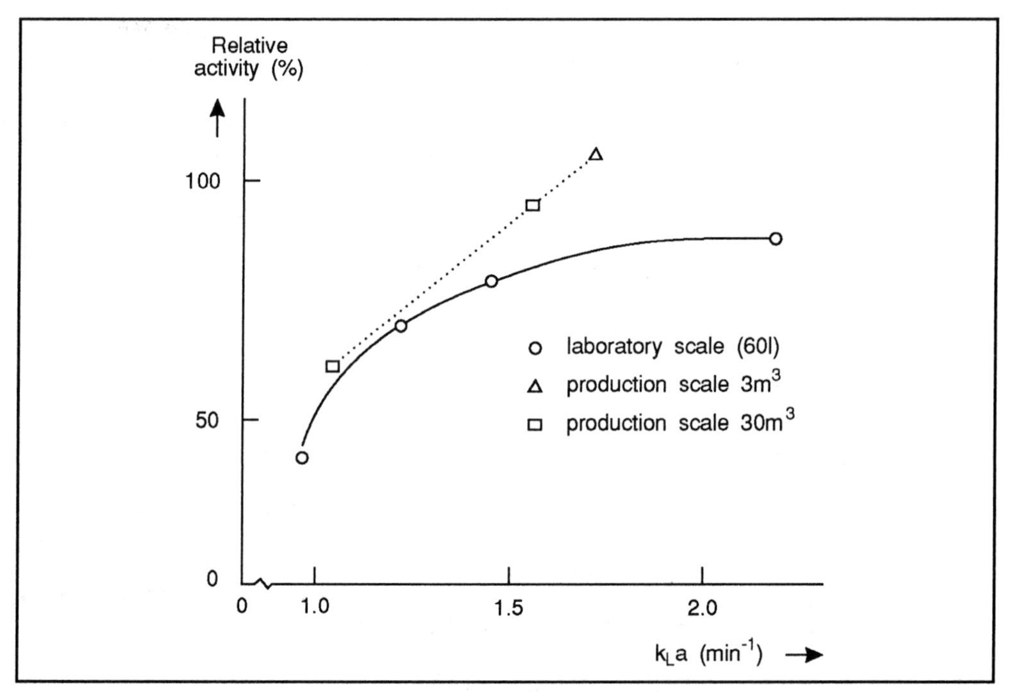

Figure 2.2 Gluco-amylase production by an *Endomyces* sp. (Redrawn from Taguchi et al, 1968 J Ferment Technol 48, 814-826).

The data described in Table 2.3 and Figure 2.2 show us that a straightforward scale up does not give the same relationship between k_La and relative activity. For example a straightforward geometric scale up from pilot to production scale would require a stirrer with a diameter of 0.79m. Applying an agitation rate of 150 rpm and an aeration rate of 0.33 vvm results in a k_La of 1.14 min^{-1} and a low (61%) product yield. Increase of the stirrer diameter at the production scale (30m^3) to 0.94m resulted in an increase of k_La from 1.1 to 1.6 min^{-1}, and a significantly better yield than expected from laboratory scale data (see Table 2.3).

Π How did increasing the diameter of the stirrer from 0.79 to 0.94 m in the production scale process influence the product yield?

The answer from the bottom line of Table 2.3 is that this increased from 61% to 91% of the equivalent pilot scale operation.

relationship between product and k_La changes at different scales

However, from Figure 2.2 it can be seen that the relation between k_La and relative activity at production scale, differs from that at laboratory scale. So, in this case an over-design of the production reactor results when this is based on the laboratory scale data. In other words, experience showed that the product yield was higher relative to k_La values (above about 1.3 min^{-1}) than the equivalent laboratory scale operation. Thus by just using a simple geometric scale up there would have been a tendency to try to achieve a higher k_La value than was really necessary.

From these two examples, we should therefore realise some of the limitations of applying 'rules of thumb'. Nevertheless, these rules can be very useful and may also be applied to scale down procedures as will be discussed later.

Generally a constant value of a particular operating/equipment variable is used. Some examples of scale up based on constant operating variables, as frequently used in fermentation technology, are already given (see for example Tables 2.1 and 2.2). The properties which can be kept constant at the different scales are:

- P/V (for CSTR = continuous stirred tank reactor);
- k_La (for CSTR and bubble column);
- v_{tip} (for CSTR);
- mixing times t_m (for CSTR);
- a combination of P/V, v_{tip} and the gas flow rate ϕ_g (for CSTR).

To keep an operating variable the same at different scales, more or less empirical correlations are often used. Geometric similarity is generally implicitly assumed. If this operating variable controls the rate-determining regime at laboratory scale and also at production scale this method is valuable (see also regime analysis below). If not, (change of regime or mixed regime) problems arise.

Before reading on, attempt SAQ 2.1.

SAQ 2.1

Consider the following relationship between product yield and P/V ratio at different scales of production.

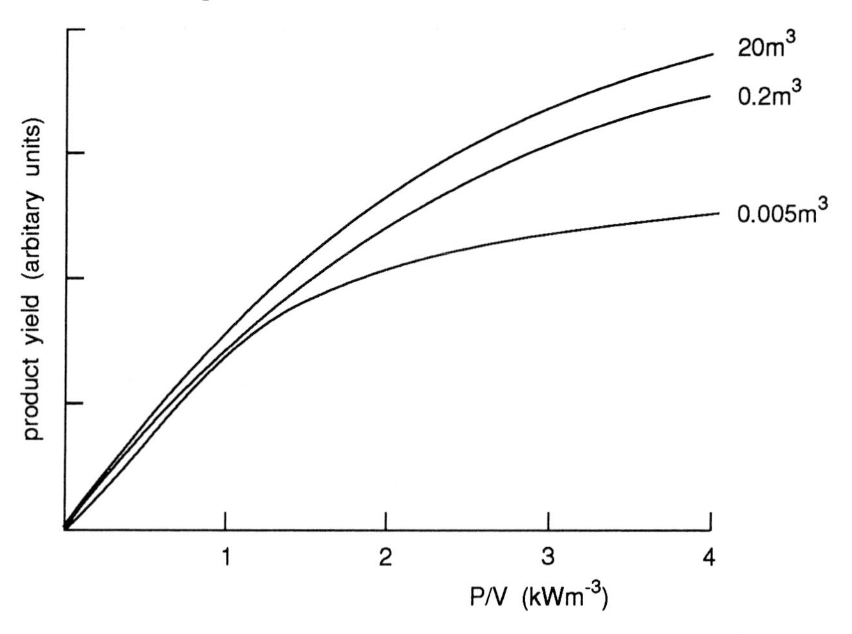

The three scales of production shown are: laboratory scale ($0.005m^3$); pilot plant ($0.2m^3$) and production scale ($20m^3$).

Which of the following statements are true and which are not?

1) At low P/V (below P/V = 1 kWm^{-3}) ratios, the product yield relative to P/V ratio is independent of scale of operation.

2) The pilot scale operation is an accurate representation of the production scale process at all P/V ratios.

3) Extrapolation of data from the laboratory scale process would tend to lead to a higher power input for the pilot plant process than is required.

4) Yields on the production scale would be lower than those estimated from laboratory scale operation.

2.3 Application of rules of thumb to scale up

In the previous section we described the properties that can be kept constant at different scales of operation. In this section we will examine each of these properties in turn.

2.3.1 Scale up of a CSTR based on constant power/volume

This criterion is often used in, for example, penicillin fermentation and other fermentations where mixing and oxygen transfer are important. As a rule of thumb, one uses a P/V ratio of 1.5-2.0 kW m^{-3} at different scales. Some data relating to penicillin production at different P/V ratios at different scales of operation are plotted in Figure 2.3.

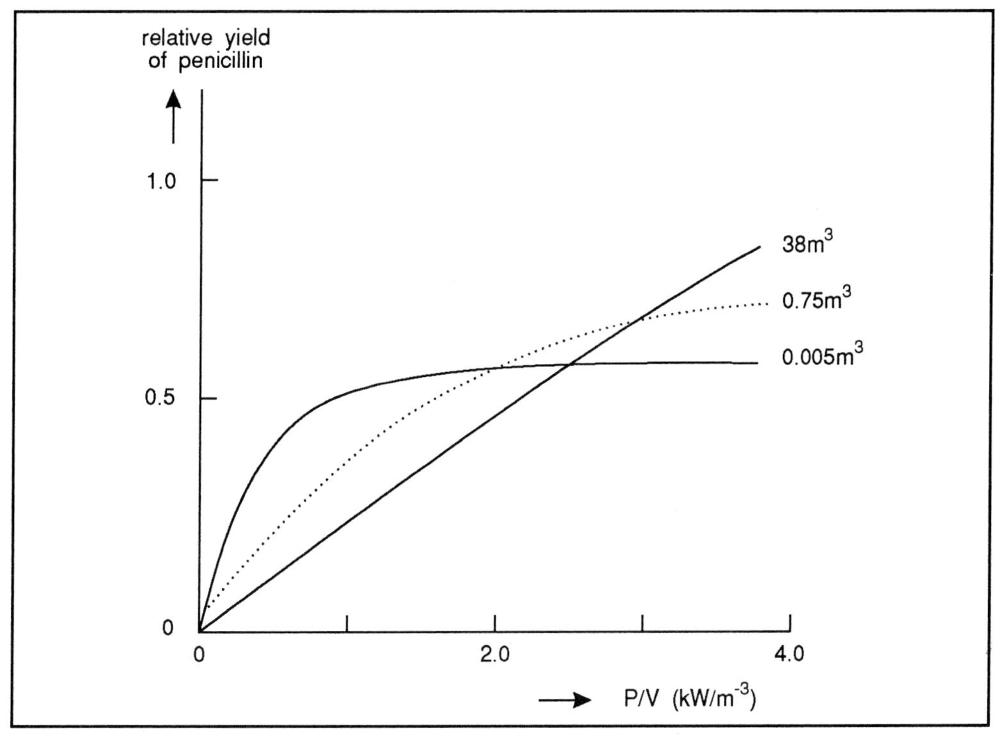

Figure 2.3 Penicillin production at different scales (after Gaden, 1961 Sci Repts 1st Super Sanita 1, 161-176).

Π The data presented in Figure 2.3 are reminiscent of that provided in SAQ 2.1. From the data presented in this figure, 1) will the yield on the production scale be higher or lower than that predicted by laboratory scale experiments, 2) at what P/V ratio does the laboratory scale experiments precisely predict the production scale process (38 m^3)?

The answer to 1) is higher at higher P/V values but lower at lower P/V values (below about 2.0-2.5 kWm^{-3}). The answer to 2) is when the P/V value is 2.5 kWm^{-3} the yield of penicillin is the same for the laboratory and production scale. Thus at this value of P/V, the laboratory scale data accurately predicts the production scale process.

In practice, the sort of data produced in Figure 2.3 show the dangers of blindly assuming that P/V has the same effects on product yield at all scales of operation. The use of laboratory scale data to design a production bioreactor may lead to over- or under design.

When P/V is used as the constant criterion, we must remember that there is a decrease in power consumption under aerated conditions. In the literature many relationships exist, but in practice the gassed power consumption is about 50% of the ungassed power consumption. It is also worthwhile knowing that correlations for gassed power consumption (P_g) and the ungassed power consumption P_S exist.

One of the more popular of these is shown in Equation 2.3.

$$\frac{P_g}{P_S} = 0.0312 \,.\, Fr^{-0.16} \,.\, Re^{0.064} \,.\, \left(\frac{\phi_g}{ND^3}\right)^{-0.38} \,.\, \left(\frac{T}{D}\right)^{0.8} \quad \text{(E - 2.3)}$$

where:

Fr = stirrer Froude number

Re = Reynolds number

N = stirrer speed

D = stirrer diameter

T = vessel diameter

ϕ_g = volumetric gas flow rate

Gassed and ungassed power consumption is discussed further in the BIOTOL text 'Operational Modes of Bioreactors' and we do not wish to pursue it at greater length here. Remember however to be careful in examining published and reported data using scale up data based on a constant power/volume. Check whether a gassed or ungassed power consumption value is being used and that there is a difference of about a factor of 2 between the two.

2.3.2 Scale up of a CSTR/bubble column based on constant k_La

We remind you that k_La is the volumetric mass transfer coefficient in which k_L = liquid film mass transfer coefficient (ms^{-1}) and a is the specific surface area based on liquid volume (m^2m^{-3}) and thus k_La has the units of s^{-1}, k_La^{20} designates the volumetric mass transfer at 20°C.

Because of the high oxygen consumption of many fermentation processes, a constant k_La at different scales is often used as scale up criterion. For example the following authors reported the use of this criterion.

Authors	Product	Reference
Bartholomew	vitamin B12 and penicillin	Adv Appl Microbiol 2, 289 (1960)
Karow and Bartholomew	streptomycin	Agricul and Food Chem 1, 302 (1953)
Taguchi *et al*	glucoamylase	J Ferment Technol 46, 814 (1968)
Steel and Maxon	novobiocin	Biotechnol Bioeng 7, 97 (1966)

popularity of k_La as scale up criterion

Again we do not anticipate that you will need to examine each of these sources. The details are given to provide you with the opportunity to follow these up if you want to. The point we wish to make is that a constant k_La has proven to be a popular criterion for scale up calculations. In the few examples described above we see examples of antibiotics, vitamins and enzymes.

However, when scaling up a reactor based on constant k_La, one needs a correlation to calculate k_La from the operating variables. For stirred tank reactors, some correlations to calculate k_La in water without salts (coalescing media) and with salts (non-coalescing media) are given below (Equation 2.4 - 2.6).

Again we point out that a full discussion of k_La values and there relationships with power consumption (P_g), volume (V) and superficial gas velocity (v_s) are described in detail in the BIOTOL text 'Operational Modes of Bioreactors'. We do however draw your attention to the following. We can divide media into broadly two types. These are:

- coalescing;
- non-coalescing.

By coalescing media, we mean media in which air bubbles tend to coalesce. In such media, the bubbles a short distance from the sparger tend to have a similar size distribution irrespective of the sparger type. In non-coalescing media, bubbles leaving the sparger will tend to retain their size distribution. Clearly the two systems are different. However, in large reactors, bubbles have more time to coalesce before they leave the system than they do in small reactors. Therefore a system that is non-coalescing on a small scale, tends to become coalescing at a large scale.

In practice the following relationships are useful ones:

for coalescing media: $$k_La^{20} = 0.026 \cdot \left(\frac{P_g}{V}\right)^{0.4} \cdot (v_s)^{0.5} \qquad \text{(E - 2.4)}$$

(v_s = superficial gas velocity)

and

for non-coalescing media: $$k_La^{20} = 0.0016 \cdot \left(\frac{P_g}{V}\right)^{0.7} \cdot (v_s)^{0.2} \qquad \text{(E - 2.5)}$$

or

in general: $$k_La^{20} = c \cdot \left(\frac{P_g}{V}\right)^{a} \cdot (v_s)^{b} \qquad \text{(E - 2.6)}$$

(Note k_La^{20} refers to volumetric mass transfer at 20°C not k_La to the power of 20!).

These equations have been derived from a great deal of experimental data and literature values. Nevertheless the parameters a, b and c in Equation 2.6, are scale dependent. This is clearly demonstrated by Table 2.4. For constant values of v_s, scale up based on constant power input and based on constant k_La should coincide but, because the exponents of Equation 2.6 vary at different scales, blindly following Equations 2.4 and 2.5 at all scales of operation, is clearly dangerous. To be accurate, using the constant k_La approach, we need reliable data for the exponents a and b at the scale of operation we are designing.

vessel size (l)	a	b
5	0.95	0.67
500	0.6-0.7	0.67
50000	0.4-0.5	0.50

Table 2.4 Exponents of Equation 2.6 at different scales (data after Bartholomew, 1960, Adv Appl Microbiol 2, 289). See text for details.

Π Compare the data in Table 2.4 and Equations 2.4 and 2.5. Is the following statement likely to be true? Equation 2.4 relates to laboratory scale operation.

Careful examination of Equation 2.4 shows the exponents a and b to be equal to 0.4 and 0.5 respectively. These are consistent with the data presented in Table 2.4 for a vessel of 50 000 l. Thus the statement appears to be incorrect and Equation 2.4 appears to relate to production scale operation. However, Equation 2.4 was found for volumes of 0.002 to 2.6 m^3 and are therefore relevant to laboratory scale operations. The results of Bartholomew show that for larger vessel sizes, a and b approach the values for calescing systems. Since exponents a and b are 0.4 and 0.5 this is consistent with exponents for coalescing media.

We might anticipate that k_La values will be influenced by the presence of antifoam agents and by medium viscosity. There are many literature correlations between k_La and these parameters. One of the most useful is to relate k_La to power input and kinematic viscosity.

In viscous fermentation broths, the dynamic viscosity η is introduced in the k_La correlations.

$$k_La = c \cdot \left(\frac{P_g}{V}\right)^a \cdot (v_s)^b \cdot (\eta)^{-0.86} \qquad \text{(E - 2.7)}$$

Although we headed this section as the scale up of a CSTR at constant k_La values, since k_La values can be described for bubble column reactors, this approach can also be used for these types of reactors.

For bubble column reactors a great number of correlations are available to estimate the k_La from the operating variables. For example, for coarse bubble systems the following equation is valid in coalescing and non-coalescing media.

$$k_La^{20} = 0.32 \cdot (v_s)^{0.7} \qquad \text{(E - 2.8)}$$

SAQ 2.2

1) We have a large volume reactor ($50m^3$) into which we are pumping air at a superficial gas velocity (v_s) of x ms^{-1}. If we reduce the gas velocity to $\frac{x}{4}$, we determine that the k_La^{20} value for the vessel is reduced by 50%. Is the medium in the vessel a coalescing or non-coalescing one? (Assume all other parameters remain constant).

2) We have a culture of a micro-organism capable of breaking down polymeric organic molecules. At the beginning of culture, we obtain a value of k_La for the culture. Some time later we determine that the k_La value has dramatically increased despite the fact that the power input and the superficial gas velocity have been kept constant. What is the most likely explanation of this observation?

2.3.3 Scale up of a CSTR based on constant tip speed

In viscous mycelial fermentations, a constant stirrer tip speed is often used as a scale up criterion, because of the shear sensitivity of the micro-organism used in this type of fermentation. The influence of scale up on shear is not clear. An increased shear can have negative effects (damage to cells) as well as positive effects (decrease of floc size). Under conditions of laminar flow the shear stress is proportional to ηN (η is the viscosity coefficient and N is the stirrer speed). At production scale laminar flow is not likely to occur, but at laboratory scale, when working with viscous broth, laminar flow conditions are more common than is often realised. Scale up then results in a change of flow regime which is rather unpleasant if one is not prepared for it. Under conditions of turbulent flow the shear stress is proportional to $\rho\,(ND)^2$, where N is the stirrer speed, D is the stirrer diameter and ρ is the density of the liquid. This results in constant values of ND being used as a scale up criterion. If not only the turbulent shear stress is important but also the frequency of exposure to this shear stress (due to fatigue phenomena or time for flocs to grow) one should expect $\rho\,(ND)^2N$ to be important. This has the same effect as keeping P/V constant. Some authors note that not only stirrer shear, but also the stirrer pumping capacity is important. They suggest stirrer shear to be divided by stirrer flow ($N^2D^2/ND^3 = N/D$) as a criterion in situations where mycelia tend to agglomerate. Others however, use only the stirrer shear as a criterion.

We may summarise these scale up criteria in the following way:

- ηN may be used if the shear stress is important in conditions of laminar flow;
- $\rho\,(ND)^2$ may be used as the scale up criterion if the turbulent shear stress is important;
- $\rho\,(ND)^2N$ may be used as the scale up criterion if the turbulent shear stress and frequency of exposure to this shear stress are important;
- N/D may be used if stirrer shear and stirrer pumping capacity are important.

2.3.4 Scale up of a CSTR based on equal mixing times

Basically large scale stirred tank reactors (with a volume larger than 5 m^3) are poorly mixed compared to small scale reactors. This is a common cause of a changing of regime (kinetic regime at model scale, transport regime at prototype scale). This can give problems for mass and heat transfer, especially in viscous broths. As an example we can use mixing in xanthan fermentation, in which the rheology of the broth changes during fermentation. To estimate the mixing time, many correlations can be used. From an energy point of view the following equation can be derived for aerated systems:

$$N \cdot t_m = 0.6 \cdot \frac{(T/D^3) \cdot (H/T)}{3\sqrt{N_P\,(H_R/D)^2}} \qquad (E-2.9)$$

in which N_P is the power number of the stirrer; H the liquid height; H_R the stirrer height. T = vessel diameter; D = stirrer diameter; N = stirrer speed; t_m = mixing time.

This type of relationship between mixing time and system parameters is derived in the BIOTOL text 'Operational Modes of Bioreactors' so we will not derive it again here. We will however comment on its general application in scale up.

Some experimental data for mixing times for different sizes of reactors used in penicillin fermentation are given in Table 2.5. In the table, these figures are compared with those obtained from Equation 2.9.

V_l (m^3)	T (m)	P (kW)	P/V (kW/m^3)	t_m (s) measured	t_m (s) calculated (from E - 2.9)
1.4	1.1	3.8	2.7	29	12
45.0	3.5	120.0	2.7	67	26
190.0	4.4	240.0	1.3	119	31

Table 2.5 Mixing times in different sizes of vessels (data from Jansen, Slott and Gurther, 1978 Proc 1st Eur Conf on Biotechnology, Intertaken Switzerland). See text for details. V_l = volume of liquid; T = vessel diameter; P = power input; t_m = mixing time.

Π From the data presented in Table 2.5, do you think Equation 2.9 is a good relationship to use in scale up.

unreliability of using constant mixing times as the scale up criterion

Your probably concluded that this was not a good relationship. The t_m calculated using this relationship consistently underestimated the real t_m. Thus in scale up terms, the use of this relationship to produce a larger system with the same t_m using various configurations of T, D, H, H_R and N_P would lead to error. Our conclusion is that use of equal mixing times, determined from theoretical considerations as the criterion used in scale up is not very reliable.

The difficulties encountered in this approach have a number of possible explanations. First, there is no generally accepted criterion for what is meant by 'being well mixed'. Secondly experience shows that aerated conditions increase mixing times as does broth viscosity.

The differences between the measured and calculated values of t_m reported in Table 2.5 suggest also that the ratio H/T has a greater effect than merely a straightforward proportional one. Thus for H/T = 2.5, the value used in Table 2.5, the t_m may be increased by a factor greater than the 2.5 that is anticipated by Equation 2.9. Thus we may summarise that the observed discrepancy between measured and calculated values may be explained by:

- when H/T = 2.5 then t_m increases more than the 2.5 predicted by Equation 2.9;
- the measurements are made under production conditions, so the viscosity of the broth can influence the mixing time;
- aerated conditions increase the mixing time;
- different criteria for being well mixed.

Clearly much more work needs to be done in this area if equal mixing times (t_m) is to be a truly effective criterion for scale up.

2.3.5 Combination of different operating variables for scale up of a CSTR

Another possibility is to keep a combination of different operating variables at the same value at the different scales. Same tip speed, power/volume ratio and aeration rate at different scales is an often used combination. The following values are quite common:

- v_{tip} larger than 3 m s^{-1};
- P/V about 2 kW m^{-3};
- ϕ_g about 0.5 vvm.

We will not comment further on this strategy at this stage.

2.3.6 Comparison of the scale up criteria

Π In order to make a comparison of the criteria used for scale up, it is a good idea to make a summary of the relationships described above. We have done this for you in Table 2.6. Copy this out onto a separate sheet. It will be a useful form of revision and you will be able to refer to the sheet during the subsequent discussions. You may notice that we have added a few additional relationships to those discussed in the text.

Power input	$P_s = 6\,\rho\,N^3D^5$	(ungassed)
	$P_g = 0.4\,P_s$	(gassed)
	$P_g = 0.0312\,P_s \,.\, Fr^{-0.16} \,.\, (Re)^{0.064} \,.\, \left(\frac{\phi_g}{ND^3}\right)^{-0.38} \,.\, \left(\frac{T}{D}\right)^{0.8}$	(gassed)
Mass transfer	$k_La^{20} = 0.026\left(\frac{P_g}{V}\right)^{0.4} \,.\, (v_s)^{0.5}$	(coalescing system)
	$k_La^{20} = 0.016\left(\frac{P_g}{V}\right)^{0.7} (v_s)^{0.2}$	(non-coalescing system)
Mixing circulation	$Nt_m = \frac{0.6\,(T/D^3)\,(H/T)}{\sqrt[3]{N_P\,(H_R/D)^2}}$	($Re > 5 \times 10^3$)
Shear	related to ηN	(laminar flow)
	related to $\rho(ND)^2$	(turbulent flow)
	related to $\rho(ND)^2N \equiv P/V$	(turbulent flow)
	related to N/D	(turbulent flow, pumping capacity important)
Some common values of process variables at production scale are:		
	P/V	2 - 4 kWm^{-3}
	ϕ_g/ND^3	0.1 - 0.15
	v_{tip}	greater than 3m s^{-1}

Table 2.6 Summary of the different scale-up criteria used for a CSTR (see text for further discussion).

Now let us make a comparison of these scale up criteria. Let us assume that for a 10 litre vessel the optimised conditions are N = 500 rpm, gas flow rate = 1 $m^3m^{-3}min^{-1}$. If we scale up this process for a 10 m^3 vessel and assume geometrically similar vessels, we can calculate the required stirrer speed. This we have done and reported the values in Table 2.7. Note that if we use a constant power/volume for a gassed system, we calculate that the required stirrer speed is 85 rpm. If on the other hand we use constant k_La as the criterion, then the calculated required stirrer speed is 79 rpm.

method		**N (rpm) for 10 m^3**
constant power/volume:		
	non gassed	107
	gassed	85
constant k_La		79
constant shear (ND)		50
equal mixing times		1260

Table 2.7 Comparison of scale-up methods. The optimum conditions for a 10 l vessel are assumed to be N = 500 rpm, gas flow rate = 1 vvm. Data from Wang 1979 Fermentation and Enzyme Technology J Wiley and Son. See text for details.

Π Which of the calculated values of N reported in Table 2.7 looks likely to be a gross over-estimation.

The most likely one appears to be the one calculated on equal mixing times. Those calculated on constant P/V, k_La and shear give comparable values. That based on equal mixing times give values over ten times greater. Clearly if we used this value, we would greatly increase the shear within the vessel (by $\frac{1260}{50}$ times); this may cause considerable damage to the culture.

Π From the data presented in Table 2.7 and the relationships described in Table 2.6, calculate the relative increase in power input required for scale up based on equal mixing times in an ungassed system to that based on a constant P/V ratio as scale up criterion for the 10 m^3 vessel described above. This is quite a complex little problem so let us work through it in a step wise manner.

We know from Table 2.6 that the power input P_s for an ungassed system is given by:

$$P_s = 6\ \rho N^3 D^5$$

To keep the $\frac{P_s}{V}$ ratio the same in the $10m^3$ vessel as was provided in the 10 l vessel, N = 107 rpm (Table 2.7). But if we wish to keep the mixing times equal, then the value of N = 1260 rpm (Table 2.7).

But the dimensions of the vessel and fluid in vessel are the same in each case (that is, ρ and D are the same in each case).

So the power input when N = 107 is $P_s = 6\rho\ (107)^3 D^5$ and the power input when N = 1260 is $P_s = 6\rho\ (1260)^3 D^5$

Thus to keep the mixing times equal we would have to increase the power input by a factor of $\frac{(1260)^3}{(107)^3}$ = approximately 1632.

We return to a point we made implicitly in the text. Scale up using constant P/V, k_La or equal mixing times does not include consideration of the effects of the varying conditions on the micro-organisms. In the example discussed above, the use of equal mixing times to calculate scale up parameters would lead to a relatively large increase in shear force and this could damage the cells considerable. Likewise using constant P/V for scale up will inevitably mean that k_La will change and this will mean that the transport of gases into the system will change. This too may have an impact on the micro-organisms being cultivated.

We complete this section by briefly considering scale up criteria for bubble columns.

The different scale up criteria for a bubble column are summarised in Table 2.8. The various criteria listed in Table 2.8 have their analogies in CSTR systems. You should note however that with bubble columns almost all studies on scale up use constant k_La as the criterion. Thus the important criteria to use are the ones cited under mass transfer in Table 2.8.

Power input	$P = v_s \rho_1 g V$
	$P = \frac{\phi_{m,gas} RT}{M_{gas}} \ln \frac{p_1}{p_2}$
Mass transfer	$k_La = 0.32 (v_s)^{0.7}$
	0.55% decrease of oxygen concentration in air per metre column (for coalescing systems)
	For non-coalescing systems k_La depends entirely on the initial bubble size
Mixing	$t_m = H^2 / De$ with $De = 0.35 (v_s \,.\, g \,.\, T^4)^{\frac{1}{3}}$
Shear	In bubble columns shear is usually not important

Table 2.8 Criteria used in the scale-up of bubble columns (P = power input, v_s = superficial gas velocity, ρ_1 = density of liquid, g = gravitational force, V = volume $\phi_{m,\,gas}$ = mass transfer of gas, R = universal gas constant, T = temperature (°K), p_1 and p_2 = pressures at the bottom and top of the column); M_{gas} = relative molecular mass of gas. ***De*** = effective diffusion coefficient.

Π If we want to have a bubble column system which has a decrease in oxygen concentration in air from 20% to 14.5%, how long does the column need to be? (Use Table 2.8 to help you). Assume the system is a coalescing one.

Experimentally we can show that the decrease in oxygen concentration in air is 0.55% per meter column. Thus to decrease the % oxygen in air from 20 to 14.5% we would need $\frac{20-14.5}{0.55}$ m column = 10 m column.

SAQ 2.3

1) A general relationship between power input for an ungassed CSTR (P_s) and for a gassed CSTR (P_g) has been established as $P_g = \alpha P_s$ where α is a coefficient. Choose the approximate value of α from the following list:

   0.04, 0.4, 4, 40, 4×10^3

2) In CSTRs the relationship between k_La, $\frac{P_g}{V}$ and v_s has been shown to be:

$$k_La = c\left(\frac{P_g}{V}\right)^a (v_s)^b$$

   From the following table of values, choose the boxes which give the combination of values for a) and b) which are found on small scale:

   i) coalescing systems;

   ii) non-coalescing systems.

		b)								
		0.1	0.2	0.3	0.5	0.6	0.7	0.8	0.9	1.0
	0.1									
	0.2									
	0.3									
	0.4									
a)	0.5									
	0.6									
	0.7									
	0.8									
	0.9									
	1.0									

   What would be the values of a and b for large scale operations?

3) Which of the following represents a common value of P/V ratios used in large scale CSTRs:

   2 Wm^{-3}, 20 Wm^{-3}, 2 kWm^{-3}, 20 k Wm^{-3}

4) In experiments with CSTRs calculated mixing times (t_m) are often longer/shorter than the measured t_m values (select the appropriate option).

2.4 Dimensional analysis

2.4.1 Introduction

So far we have examined the application of 'rules of thumb' to scale up problems. We now turn our attention to another approach, namely that which uses dimensional analysis. This is a widespread method, commonly used in the scale up of chemical engineering problems, which can be very useful for scale up of microbial processes. Dimensional analysis is a technique by which dimensionless groups of parameters, for example the Fr and Re number, can be obtained from the dimensionless momentum, mass, and heat balances, as well as their boundary and initial conditions in the dimensionless form. The basic techniques employed in dimensional analysis are described in detail in the BIOTOL text 'Bioprocess Technology: Modelling and Transport Phenomena'. Dimensional analysis is however of such importance and so open to misinterpretation that we consider it essential to review this approach in order to show how it may be employed in scale up problems.

The technique of dimensional analysis is driven by the need for dimensional consistency and the constraints this places on the functional relationship between variables. Essentially what the technique does is to allow us to group a number of variables in a problem to form dimensionless groups. In general, dimensionless numbers are ratios (quotients) between two fundamental properties. For example Reynolds number (Re) is the ratio between inertia forces and viscous forces.

$$Re = \frac{\text{inertia forces}}{\text{viscous forces}}$$

If we convert this to the mathematical form then:

$$Re = \frac{\rho v D}{\eta} \text{ for flow through a tube or}$$

$$Re = \frac{\rho N D^2}{\eta} \text{ for a stirred vessel.}$$

If we substitute in the units for ρ, v, D, η and N into these equations, they will cancel each other out to produce a dimensionless number, in this case Reynolds number.

examples of dimensionless numbers

Many dimensionless numbers have been determined. These are summarised in Table 2.9. This is quite an extensive list. Many of the numbers included in this list you will have already met. Examine this table carefully and see if you can answer the following questions.

Π 1) Which dimensionless number describes the ratio of mass transfer by convection and mass transfer by diffusion? 2) Which dimensionless number provides the ratio between the total heat transfer and the heat transfer by conduction? 3) Which dimensionless number gives the ratio between total mass transfer and mass transfer by diffusion?

The answers to these questions are 1) Péclet; 2) Nusselt and 3) Sherwood. The key point to remember about the application of dimensionless numbers in scale up is that each of

Category	Number	Definition	Formula
Momentum:	Reynolds	$\frac{\text{inertia forces}}{\text{viscous forces}}$	$Re = \frac{\rho v D}{\eta} \left(Re^{b)} = \frac{\rho N D^2}{\eta} \right)$
	Froude	$\frac{\text{inertia forces}}{\text{graviational forces}}$	$Fr = \frac{v^2}{gL} \left(Fr^{b)} = \frac{N^2 D}{g} \right)$
	Weber	$\frac{\text{inertia forces}}{\text{surface forces}}$	$We = \frac{\rho v^2 d_p}{\sigma}$ $\left(We^{b)} = \frac{\rho N^2 D^2 d}{\sigma} \right)$
	Power number	$\frac{\text{total dessipated power}}{\text{power due to inertia}}$	$Po = \frac{P}{\rho N^3 D^5}$
Mass:	Sherwood	$\frac{\text{total mass transfer}}{\text{mass transfer by diffusion}}$	$Sh = \frac{kD}{\boldsymbol{D}}$
	Schmidt	$\left(\frac{\text{hydrodynamic boundary layer}}{\text{mass transfer boundary layer}} \right)$	$Sc = \frac{v}{\boldsymbol{D}}$
	Péclet	$\left(\frac{\text{mass transfer by convection}}{\text{mass transfer by diffusion}} \right)$	$Pe = \frac{vL}{\boldsymbol{D}}$
	Fourier	$\left(\frac{\text{process time}}{\text{diffusion time}} \right)$	$Fo = \frac{\boldsymbol{D} t}{D^2}$
	Biot	$\left(\frac{\text{external mass transfer}}{\text{internal mass transfer}} \right)$	$Bi = \frac{k d_p}{\boldsymbol{D}^{a)}}$
Heat	Nusselt	$\left(\frac{\text{total heat transfer}}{\text{heat transfer by conduction}} \right)$	$Nu = \frac{\alpha D}{\lambda}$
	Prandtl	$\left(\frac{\text{hydrodynamic boundary layer}}{\text{thermal boundary layer}} \right)^3$	$Pr = \frac{v}{a}$
	Péclet$_h$	$\left(\frac{\text{heat transfer by convection}}{\text{heat transfer by conduction}} \right)$	$Pe_h = \frac{vL}{a}$ $[a = \lambda/(\rho c_p)]$
	Fourier$_h$	$\left(\frac{\text{process time}}{\text{heat conduction time}} \right)$	$Fo_h = \frac{at}{D^2}$
	Biot$_h$	$\left(\frac{\text{external heat transfer}}{\text{internal heat transfer}} \right)$	$Bi_h = \frac{a d_p}{\lambda^{a)}}$
Chem React	Damköhler I	$\left(\frac{\text{chemical reaction rate}}{\text{mass transport by convection}} \right)$	$Da_I = \frac{rL}{vC}$
	Damköhler II	$\left(\frac{\text{chemical reaction rate}}{\text{mass transport by diffusion}} \right)$	$Da_{II} = \frac{rL^2}{\boldsymbol{D} C}$
	Thiele Modulus	$\left(\frac{\text{chemical reaction rate in particle}}{\text{diffusion in particle}} \right)^{\frac{1}{2}}$	$\phi = R\sqrt{r/\boldsymbol{D} C}$

Table 2.9 Dimensionless numbers (adapted from Kossen and Oosterhuis 1985 Biotechnology Vol 2 Ed Rehm and Reed VCH Verlagsgesellschaft Weinheim). [a] indicating that $\boldsymbol{D}$ or λ are related to the dispersed phase; [b] for stirred vessels. Subscript h refers to heat transfer. For other symbols see appendix.

these numbers measure the ratio of two fundamental parameters. In principle we can aim to keep these ratios constant as we scale up a process.

You should note that some dimensionless groups (We and Bi) typically refer to two-phase systems. Some groups appear for mass transfer as well as for heat transfer. All of them can be interpreted as a ratio of time constants. However, it is often quite difficult to obtain the dimensionless groups. There are many situations in which a microbalance equation cannot be set up.

scale up based on constant dimensionless numbers

Once the dimensionless groups have been obtained their use for the proper set up of scale up or down experiments is, at least in principle, rather simple: equal values of these groups for both the model scale and the prototype scale systems. In practice, the situation is more complicated, because often it is not possible to keep all the dimensionless groups constant during scale up.

For example, looking to a mixing problem in a stirred vessel with a homogenous fluid, the Reynolds number and the Froude number have to be kept the same at different scales.

$$Re_M = Re_P: \quad \left(\frac{\rho N D^2}{\eta}\right)_M = \left(\frac{\rho N D^2}{\eta}\right)_P \qquad \text{(E - 2.10)}$$

$$Fr_M = Fr_P: \quad \left(\frac{N^2 D}{g}\right)_M = \left(\frac{N^2 D}{g}\right)_P \qquad \text{(E - 2.11)}$$

in which subscript M means the model and subscript P the prototype, ρ the density of the liquid and g is the acceleration due to gravity. If the same liquid is used for model and prototype, the conditions ND^2 = constant and N^2D = constant cannot be matched at different scales (try it!).

In principle, one can, however, choose different liquid systems for model and prototype in order to match the equality of Re and Fr. This is the case if:

$$\frac{\rho_M}{\eta_M} = \frac{\rho_P}{\eta_P} \cdot \left(\frac{D_P}{D_M}\right)^{3/2} \qquad \text{(E - 2.12)}$$

which results in a much lower value of η_M compared with η_P if $\rho_M \sim \rho_P$. In practice in bioreactors this is not easy to realise. With biological systems we are more-or-less confined to using water and the growth of the cells is influenced by additives in the water. Thus, in these cases it is difficult to manipulate ρ_M , ρ_P , η_M and η_P.

Thus we have to modify our strategy by basically asking the question, which of the dimensionless numbers we are trying to keep constant is most important (ie reflect the rate limiting mechanism). In the case of the Reynolds and Froude numbers described above, if the vessel is fully baffled, the Froude numbers play no role, and it will be sufficient to keep the Re-number in the turbulent regime ($Re > 10^4$). So one has to determine the most important groups of parameters, neglecting the rest. However, if a change in regime takes place during scale up, the formal dimensional analysis method breaks down completely.

Another disadvantage of the dimensional analysis can be that the formal application of dimension analysis leads to technically unrealistic situations (too large power consumption, stirrer speed, etc). For heterogeneous systems another problem arises. During scale up one is usually not interested in geometric similarity for the dispersed particles present in the heterogeneous system. Generally, the size of flocs, pellets, air bubbles, droplets and crystals do not grow proportionally with the scale of the apparatus but remain constant during changes of scale. This violates the basic concept of dimensional analysis: **geometric similarity**. Finally, the choice of parameters involved in dimensional analysis is not always obvious and sometimes rather arbitrary.

By now you may be feeling a little confused. We began this section by claiming that dimensional analysis is used widely and then proceeded to highlight the difficulties and limitations of this method. The two aspects are not however incompatible. Dimensional analysis offers a considerable potential for scale up (or scale down as will be discussed later) by reducing the number of variables and the number of experiments at laboratory scale, which are necessary to predict the system behaviour at production scale. But we must be aware of its limitation. Because of its importance the generation of the dimensionless groups is discussed in the next section together with some examples.

2.4.2 Generation of the dimensionless groups

There are many ways to generate the dimensionless groups of parameters. One method has already been presented: using the dimensionless momentum, mass and heat microbalances, as well as their boundary and initial conditions in the dimensionless form. Often it is not possible to set up a microbalance. The first and main problem that has to be solved then is to determine the relevant parameters. This inventory is facilitated if different groups of parameters are distinguished:

- geometrical parameters (eg D, H, d_p)
- fluid/solid/gas properties (eg ρ, η, σ)
- process variables (eg N, P, v)
- dimensional constants (eg g, R)

where D = diameter of stirrer, H = height of vessel, d_p = diameter of particles, ρ = density; η = viscosity, N = stirrer speed, P = power input, v = velocity, g = gravitational constant, R = gass constant, σ = surface tension.

importance of experience in applying dimensional analysis

The inventory is a matter of experience and intuition. If too many parameters are collected, including non-relevant ones, too many groups are generated that have to be kept at a constant value for both the model and the prototype. If a relevant parameter is overlooked, an important group will be missing, resulting in an incomplete description of the system. In practice, non-relevant groups can be eliminated using regime analysis (we will discuss this later). Missing important groups usually are detected during experimentation when it appears not to be possible to describe the behaviour of the system with the existing groups.

Once the relevant parameters have been obtained, their collection into different dimensionless groups is a well established standard procedure. We remind you of the principles of the procedure. Essentially the procedure involves the following stages:

Stage 1 - the quantity (A) we wish to consider is chosen;

Stage 2 - the physical parameters which determine the chosen one are determined (B_1 to B_n);

Stage 3 - write this in the form $A = B_1^{\alpha} \; . \; B_2^{\beta} \; .... \; B_n^{\mu}$, where α, β etc are functions of B_1, B_2 etc;

Stage 4 - now replace A, B etc by their dimensions making sure that they are homogeneous (eg do not mix cm and m), then all the dimensions that occur are examined;

Stage 5 - use the functions α, β etc to group quantities A, B_1 to B_n into dimensionless numbers.

NB The BIOTOL text 'Bioprocess Technology: Modelling and Transport Phenomena' provides a fuller description of this procedure with full examples of how to work through this procedure.

Also important in dimensional analysis is the Buckingham II theorem. If when we have worked out the number of parameters (n) which influence the chosen quantity and these parameters contain (m) basic units, then Buckingham's theorem states that the smallest set of dimensionless groups that describes the system is n-m.

In other words if we have 5 parameters composed on 3 different units, then the smallest set of dimensionless groups will be 5 - 3 = 2.

In Stage 3 to 5 of dimensional analysis, described above, we indicated how we group the quantities into dimensionless numbers. There are a number of methods available to do this. The main ones are:

- Gauss-Jordan reduction method;
- Rayleigh's method;
- Gukhman's method.

Rayleigh's method will be demonstrated below. A variant of the method of Gukhman has been published by Quraishi and Fahidy (1981 Can J Chem Eng 59, 563-566) and the procedure will not be discussed further here. Let us now examine an example:

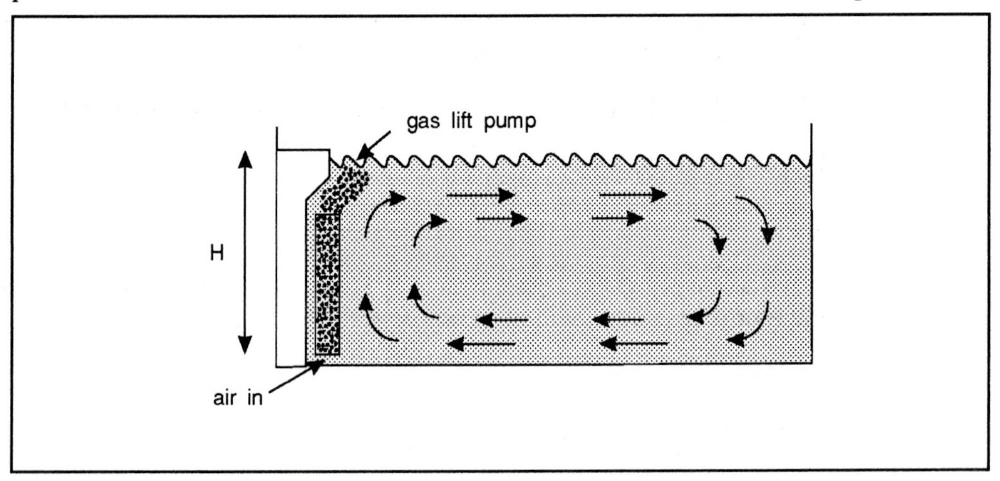

Figure 2.4 Sludge digester used in Example 1 (see text for details).

In a large sludge digester (Figure 2.4), microbial flocs have to be kept in suspension due to a circulating flow induced by gas lift pumps. A gas lift pump is a kind of a draft tube placed in the digester with a gas sparger at the bottom.

It became clear from some preliminary runs with the prototype that the pumps were not capable of maintaining the required concentration of solids. Because experiments on the prototype scale are very expensive it was decided to scale the system down and to look for a solution at the model scale.

The key to success in this approach is to decide what the relevant parameters are. After some considerable thought, the following parameters were considered relevant:

characteristic dimension	H	(m)
liquid velocity	v	($m\ s^{-1}$)
density of the liquid	ρ	($kg\ m^{-3}$)
density difference between the liquid in the tank and the gas/liquid mixture in the gas lift pump	$\Delta\rho$	($kg\ m^{-3}$)
acceleration due to gravity	g	($m\ s^{-2}$)

According to the Buckingham II theory, five parameters and three dimensional quantities should result in two dimensionless groups. The method of Rayleigh is used to obtain the groups. It is assumed that the solution of the problem can be written as:

$$v = H^{\alpha} . \rho^{\beta} . \Delta\rho^{\gamma} . g^{\delta}$$

By the use of dimensions this relation can be written as:

$$\frac{L}{T} = L^{\alpha} \left(\frac{M}{L^3}\right)^{\beta} . \left(\frac{M}{L^3}\right)^{\gamma} . \left(\frac{L}{T^2}\right)^{\delta}$$

where M is dimension of mass, L of length and T of time. Because every physical meaningful relation ought to be homogeneous in its dimensions we come to the following conclusions for the dimension M, L and T:

M: $0 = \beta + \gamma$ (because M does not appear on the left hand side of the equation)

L: $1 = \alpha - 3\beta - 3\gamma + \delta$

T: $-1 = -2\delta$

We now have four unknown powers (α, β, γ and δ). However, we have three equations, therefore, three unknown powers can be eliminated:

The first equation thus leads to: $\beta = -\gamma$

The third equation yields: $\delta = 1/2$

The second equation leads to the conclusion that $\alpha = 1/2$

The powers in the original equation can now be rewritten as:

$$v = H^{\frac{1}{2}} \cdot \rho^{-\gamma} \cdot \Delta\rho^{\gamma} \cdot g^{\frac{1}{2}}$$

We can now group together equal powers to give:

$$\frac{v}{(gH)^{0.5}} = \left(\frac{\Delta\rho}{\rho}\right)^{\gamma}$$

or more generally:

$$\frac{v}{(gH)^{0.5}} = f\left(\frac{\Delta\rho}{\rho}\right)$$

Thus we have two dimensionless numbers related to each other by a function f. The physical meaning of one of these dimensionless numbers is self evident. $\frac{\Delta\rho}{\rho}$ is the ratio of the difference in the density of the liquid and gas/liquid mixture in the gas lift pump, to the density of the liquid. But what is the physical meaning of $\frac{v}{(gH)^{0.5}}$?

manipulation of dimensionless groups

This is a lot more difficult to visualise. It is quite permissable to carry out mathematical manipulations to bring an equation into a more useful form. For example say we have the relationship:

(Dimensionless Number_1) = f (Dimensionless Number_2)

We can do such manipulations as multiply each side by the same value. For example if we multiply the equation above by Dimensionless number_2 we could get:

(Dimensionless Number_1) (Dimensionless Number_2) = f (Dimensionless $\text{Number}_2)^2$

Equally we can carry out other mathematical manipulations. This is quite important as it enables us to bring dimensionless numbers into a form that has a clear physical meaning.

Let us return to our relationship:

$$\frac{v}{(gH)^{0.5}} = f\left(\frac{\Delta\rho}{\rho}\right)$$

After some trial and error manipulations, we decide to adopt the following strategy.

To obtain groups with a clear physical meaning we take the square of the first group and multiply this with the reverse of the second group. The result is:

$$\frac{\rho\, v^2}{\Delta\rho \cdot g \cdot H} = f\left(\frac{\Delta\rho}{\rho}\right)$$

The physical interpretation of the first group is: resistance due to the flow of the circulating liquid relative to the buoyancy of the gas/liquid mixture in the gas lift pump. If both dimensionless groups are kept at a constant value the result is:

$$\left(\frac{\Delta\rho}{\rho}\right)_M = \left(\frac{\Delta\rho}{\rho}\right)_P$$

$$v_M = v_P \left(\frac{H_M}{H_P}\right)^{\frac{1}{2}}$$

This, in the gas lift pump, is determined by the gas hold-up in the pump. This gas hold-up is proportional to:

$$\frac{v_s}{v_b} = \frac{\phi_g}{H^2 v_b}$$ ϕ_g = gas flow rate, v_b = velocity of bubble, v_s = velocity of solvent

Because the rising velocity of air bubbles v_b is roughly constant for bubble diameters larger than 2 mm, this means that, in order to keep it constant, ϕ_g must be proportional to H^2 or:

$$\phi_{g,M} = \phi_{g,P} \cdot \left(\frac{H_M}{H_P}\right)^2$$ where H_M = height of model, H_P = height of prototype

Thus using dimensional analysis, it was shown that the gas flow rates at the model and prototype scales should be in the ratio of $\left(\frac{H_M}{H_P}\right)^2$

The model was built according to these scaling criteria. Improvements were found by experimentation with the model. These improvements appeared to be successful after application in the prototype.

What this example shows is that a dimensional analysis approach enabled an appropriate model to be built. This in turn enabled appropriate experiments to be conducted which were relevant to the prototype scale.

A second example of the application of dimensional analysis in scale up is described in Appendix 1. This second example uses the Gausse-Jordan reduction method to obtain the dimensionless groups. This is not a technique which is so commonly used, thus we have included it as an 'optional' case for consideration.

SAQ 2.4

It is proposed to model a prototype process involving a long tube in which cells are cultured. After some consideration the following parameters were thought to be relevant.

C_0 = concentration of the substrate at the point of entry (kg m^{-3})

D = diffusion coefficient (m^2s^{-1})

v = velocity of the liquid in the tube (ms^{-1})

C_L = the concentration of the substrate at the exit of the tube (kgm^{-3})

L = length of the tube (m)

Thus we believe that:

$C_L = f(C_0, D, v, L)$ or more specifically $C_L = C_0^{\alpha} . D^{\beta} . v^{\gamma} . L^{\delta}$

Use this information to carry out a dimensional analysis of the system to identify a) dimensionless number(s) than can be used as constant in the scaling process. Do this in a number of stages.

1) Apply Buckingham Π theorem to determine the minimum number of dimensionless numbers.

2) Use Rayleigh's method to determine this (these) dimensionless number(s).

3) Use Table 2.9, to identify the dimensionless number(s) that can be used in the scaling process.

2.5 Similarity principle

The similarity principle is used quite often in the scale up of fermentation processes, although mainly in combination with semi-fundamental, the rules of thumb or dimensional analysis methods. In the previous sections we have already seen some examples of geometric similarity of model and prototype. There are, however, also other similarity principles. In general, model and prototype are:

- geometrically similar if the geometric ratios are similar;
- hydrodynamically similar if, along with geometric similarity, the dimensionless flow profiles are also similar;
- thermally similar, if, together with geometric similarity and hydrodynamic similarity, the dimensionless temperature profiles are also similar;
- chemically similar if the dimensionless concentration profiles are similar and if the model and prototype are similar in all other aspects.

From the above it is clear that the method involves the use of balances for momentum, heat and mass in dimensionless forms. Likewise their boundary and initial conditions are also described in dimensionless forms. Thus, if the model and prototype are geometrically similar and if the dimensionless groups and boundary conditions are the same for the model and the prototype, then the solution of the balance equation in this dimensionless form is the same for model and prototype. However the non-dimensionless profile can be quite different.

Let us illustrate this by the mass balance for a plug flow reactor, fitted with dispersion but without accumulation (ie dC/dt = 0).

Let us assume that the length of the reactor is L and x is a position along the reactor. Thus:

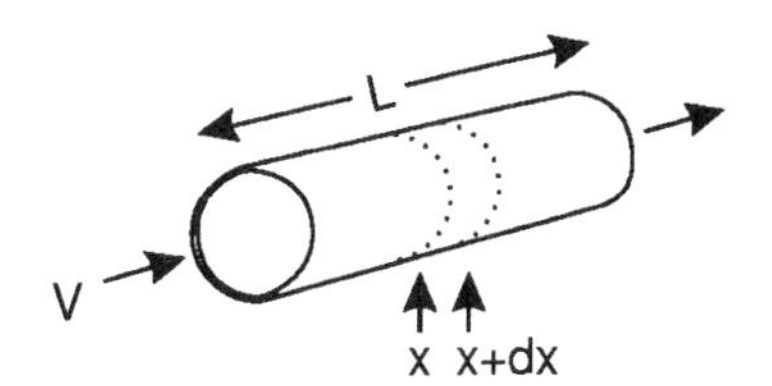

In principle we can identify the position along the reactor in dimensional units. For example if L = 1m and x was halfway along the tube, x = 0.5m. Equally, we can identify the position x in dimensionless units simply by dividing by L, ie $\frac{x}{L} = 0.5$. Thus we have identified the position of x as halfway from one end and this identification is in dimensionless units.

Now let us return to the mass balance for the plug flow reactor. We can write:

$$D \cdot \frac{d^2C}{dx^2} - v \cdot \frac{dC}{dx} + kC = 0 \qquad \text{(E - 2.13)}$$

where C = concentration, v = liquid velocity, *D* = diffusion coefficient.

This equation can be written in dimensionless form as:

$$\frac{D\,C_0}{L^2} \cdot \frac{d^2\,(C/C_0)}{d\,(x/L)^2} - \frac{v\,.\,C_0}{L} \cdot \frac{d\,(C/C_0)}{d\,(x/L)} + r\,.\,C_0\,(C/C_0) = 0 \qquad \text{(E - 2.14)}$$

or:

$$\frac{D}{r\,.\,L^2} \cdot \frac{d^2\,(C/C_0)}{d\,(x/L)^2} - \frac{v}{r\,.\,L} \cdot \frac{d\,(C/C_0)}{d\,(x/L)} + C/C_0 = 0 \qquad \text{(E - 2.15)}$$

The dimensionless groups involved in the equation are:

$$Da = \frac{r\,.\,L^2}{D} = \frac{\text{diffusion time}}{\text{conversion time}} \left(\text{or} \quad \frac{\text{chemical reaction rate}}{\text{diffusion rate}} \right)$$

and

$$\frac{r\,.\,L}{v} = \frac{\text{residence time}}{\text{conversion time}} \left(\text{or} \quad \frac{\text{chemical reaction rate}}{\text{convection rate}} \right)$$ - see Table 2.9

If, therefore, model and prototype are geometrically similar and if the dimensionless groups and the dimensionless boundary conditions are the same for model and prototype, then the solution of the balance equation in its dimensionless form is also the same for model and prototype. However, the non-dimensionless profiles can be quite different. This is shown in Figure 2.5 where the dimensionless profile is given together with the non-dimensionless profiles of model and prototype. Furthermore, keeping the model and prototype similar can sometimes lead to complicated balance equations which may be difficult to solve and thus it becomes almost impossible to derive dimensionless profiles.

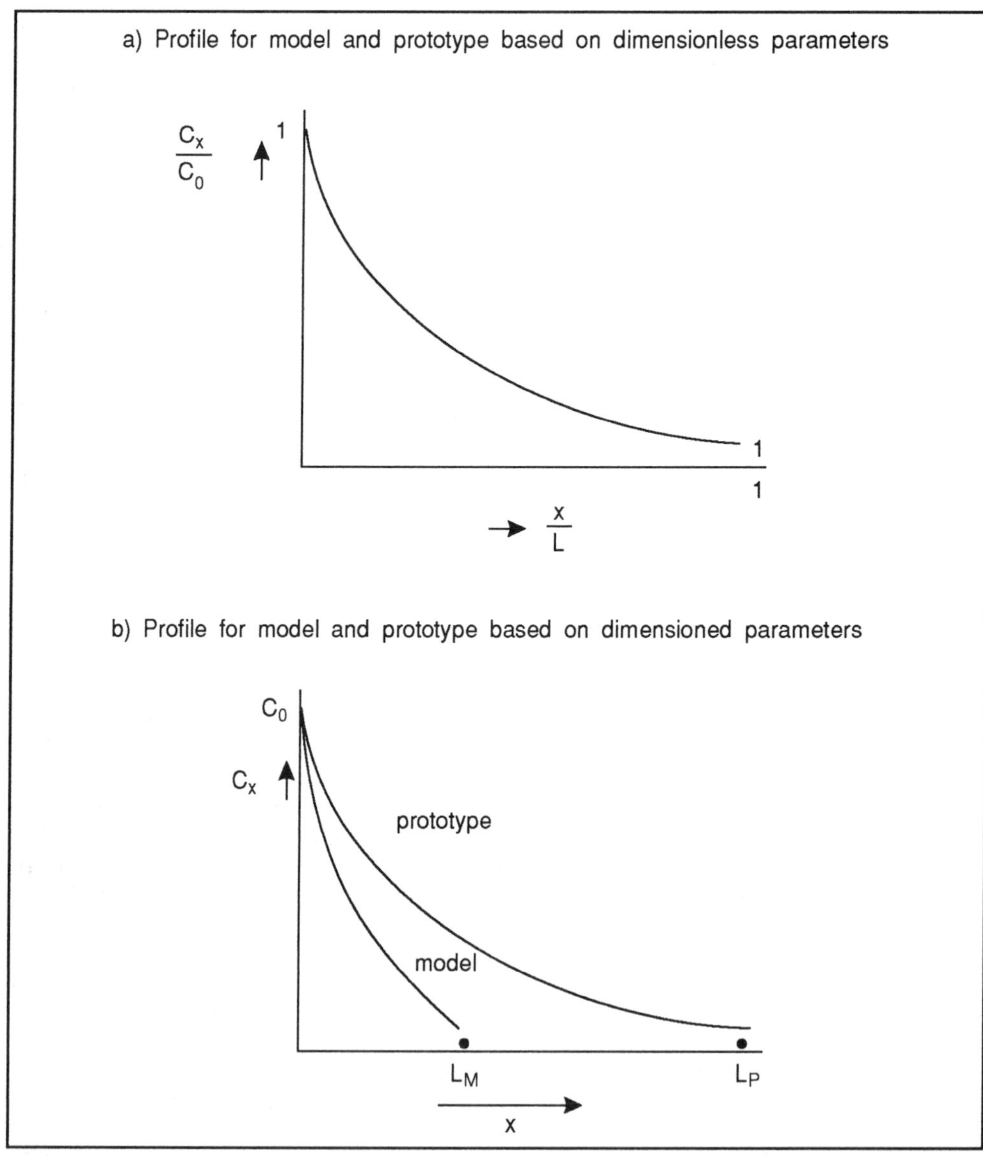

Figure 2.5 Comparison of the profiles of concentration in a plug flow reactor based on a) dimensionless and b) dimensioned parameters (C_0 = inlet concentration, C_x = concentration at position x, L = length of reactor, x = position along the reactor, subscript M denotes model, P denotes prototype). See text for details.

SAQ 2.5

We are attempting to use a model to predict the temperature of water in a pipe. The model tube is 1m long and water flows along the pipe at 0.1 ms^{-1}. The temperature of the water in the pipe has been measured at fixed points and gave the following results.

Distance from inlet (x)	Temperature (°C)
0 (ie inlet)	$T_0 = 60$
25 cm	$T_{25} = 50$
50 cm	$T_{50} = 42$
75 cm	$T_{75} = 35$
100 cm (ie outlet)	$T_{100} = 29$

1) The proposed prototype is to be 100m long and it is proposed that the velocity of water flowing in the prototype is to be 10ms^{-1}. Use a dimensionless model to predict the temperature of the water leaving this pipe, if the inlet water is at 60°C. Assume that the temperature profile is not influenced by the diameter of the pipe, and that there is no difference in the materials used in the manufacture of the pipe, but that the temperature profile is influenced by the velocity of water in the pipe.

2) Plot graphs of the actual (dimensioned) temperature profiles in the model and prototype pipes.

2.6 Environmental approach

So far the approaches we have described have paid rather little attention to the biological system being used in the bioreactor. We know however that such systems are 'delicate' and their behaviour in terms of growth, product yield etc is greatly influenced by their environment. The environmental approach pays more attention to the conditions the micro-organisms will meet on their way through the reactor. It is based on small scale simulation of the environmental conditions in the production scale fermentor. The micro-environment, which can influence the process, can be summarised as follows:

- chemical variables - carbon, nitrogen, phosphorous source, oxygen concentration, product formation, other (like pH, precursors, antifoam);
- physical variables - temperature, viscosity, non-aqueous liquid or solid substrate/product distribution, power input, shear, microbial morphology.

The backbone of this approach is trying to keep all these variables and their fluctuations, which can influence the microbial system, the same at the different scales. Because of the fluctuations of the micro-environment on the production scale, it will be impossible to keep all aspects of the micro-environment the same at the different scales. This approach is therefore difficult to apply and we will not develop it further at this stage

(Further details of this approach can be obtained from Young, 1979, Ann of the New York Acad of Sci 326, 165-180).

A much more realistic approach is to first determine the rate limiting step. From this, the production reactor is designed to optimise the rate limiting step. Although each system will present its own particular problems we can idealise the approach used by the following stages:

- establishment of the rate-limiting biological step;
- establishment of the relation between the physiology of the micro-organism and the external environment;
- establishment of the relationship between the operating and equipment variables.

Clearly this approach demands close liaison between biologists and technologists.

2.7 Regime analysis

2.7.1 Introduction

characteristic times

The regime concept was introduced by Johnstone and Thring (1975) and is a necessary ingredient for scale down of fermentation processes. Regime analysis is based on the use of so-called characteristic times. Characteristic time is a measure of the rate of a mechanism or sub-process and can be considered as the time needed by that mechanism to reach a certain percentage of its final value after a (step-wise) change. Each mechanism or sub-process can be characterised by such a value (see Figure 2.6). A low value of a characteristic time means a fast mechanism; a high value means a slow mechanism. The use of these characteristic times can also give insight into the complexity of the process. When different time constants are of the same order of magnitude, we are speaking of a mixed regime. In that case scale up or down of the process will cause problems. On the other hand, when scale translation is performed with respect to one particular regime that is rate limiting at laboratory scale, it is possible that another regime is rate determining at production scale. Then we have a so-called change of regime. This also can cause problems.

In the literature terms like time constant, process time (constant) and relaxation time are also used instead of characteristic time. The term time constant is commonly used for first order processes only. These times can be measured, calculated or estimated. For estimation the following relation is used:

$$t_{char} = \text{capacity} / \text{flow} \qquad \text{(E - 2.16)}$$

By listing all sub-processes and calculating (determination or estimation) the characteristic times, the rate-limiting mechanisms can be determined. Such a listing not only has to be done for the final production scale which one has in mind or which is already available; the same inventarisation has to be done for the laboratory and eventually the pilot plant scale, to predict the possibility of the regime changing on scale up or scale down. In many fermentation systems problems in scale up can be explained by change of regime. From the knowledge originating from the regime analysis, it can be concluded which mechanisms need further investigation on a small scale.

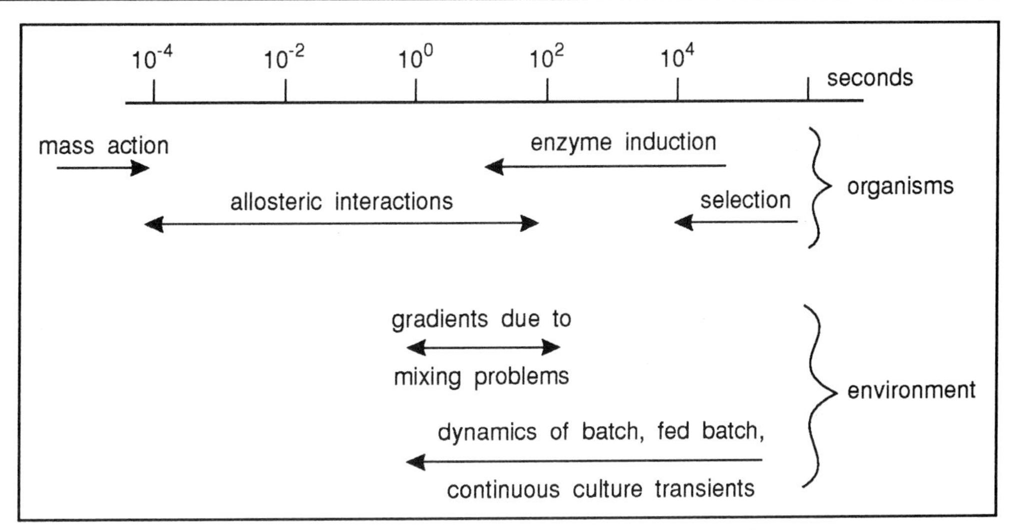

Figure 2.6 Typical time constants in interactions between organisms and the environment.

A drawback of this approach with regime analysis can be the lack of information of production scale systems. In many cases regime analysis is only possible using non accurate model equations which are not validated by large scale experimental information. However, a first estimate of the value is normally sufficient enough to identify the rate-limiting mechanisms and to predict whether there will be a change in regime if the process is scaled up. For the first estimate the various equations as presented below can be used. The characteristic times have been separated for transport and conversion phenomena. The characteristic times for transport phenomena are dependent on reactor-type (CSTR/bubble column) while those for conversion phenomena are found to be independent of reactor type.

SAQ 2.6

Choose the correct word from the list provided below to complete the following.

1) A [ ] regime is one in which the time constants are of the same order.

2) A [ ] of regime is the phrase used when different regimes are rate determining at laboratory and production scales.

3) [ ] time is a measure of the rate of a mechanism.

4) [ ] phenomena are dependent on reactor type; [ ] phenomena are not.

Word list: conversion, mixed, homogenous, mixture, change, calculation, relaxation, process, characteristic, transport, two phase.

2.7.2 Characteristic times for transport phenomena

Oxygen transfer

The characteristic time for oxygen transfer t_{OT} from gas to liquid is defined by the reciprocal of k_La:

$t_{OT} = 1/k_La$ (E - 2.17)

For design purposes k_La can be calculated for stirred vessels according to the relationships:

$$k_La^{20} = 0.026 \left(\frac{P_g}{V}\right)^{0.4} . (v_s)^{0.5}$$ (coalescing medium)

and $$k_La^{20} = 0.0016 \left(\frac{P_g}{V}\right)^{0.7} . (v_s)^{0.2}$$ (non-coalescing medium)

(see Equations 2.4 and 2.5)

For bubble columns the relationship

$k_La^{20} = 0.32\,(v_s)^{0.7}$ can be used (see Equation 2.8).

Circulation time

The characteristic time for mixing t_m of the gas-liquid dispersion is expressed by the circulation time t_c. In the case of a stirred tank t_m equals, approximately 4 times t_c ; t_c can be calculated from:

$$t_c = \frac{V}{\phi_P}$$ (E - 2.18)

in which V is the volume of the liquid and in which the pumping capacity ϕ_P of a turbine type of stirrer can be calculated from:

$\phi_P = 0.75\,N . D^3$ (E - 2.19)

where N = stirrer speed, D = stirrer diameter.

In a CSTR with more than one stirrer, the circulation time has to be calculated for each one stirrer compartment. In that case the compartment volume V as used to calculate the circulation time can be estimated from:

$$V = \frac{\pi}{4} . T^2 . T$$ (E - 2.20)

where T is the vessel diameter

(Note calculation of compartment volume is dealt with in the BIOTOL text 'Operational Modes of Bioreactors').

In the case of a bubble column mixing time (t_m) equals 2 to 5 times t_c ; t_c can be calculated from:

$$t_c = \frac{H}{v_{lc}}$$ (E - 2.21)

in which v_{lc} is the liquid circulation velocity.

Gas residence time

The gas residence time of the gas bubbles t_G is derived from the gas hold-up in the reactor and is in the case of a CSTR:

$$t_g = (1 - \varepsilon) \cdot \frac{V}{\phi_g} \qquad \text{(E - 2.22)}$$

where ε = gas hold up, ϕ_g = gas flow rate.

In the case of a bubble column:

$$t_g = (1 - \varepsilon) \cdot \frac{H}{v_s} \qquad \text{(E - 2.23)}$$

Π Can you prove that Equations 2.22 and 2.23 are in fact the same equation?

Since $v_s = \frac{\phi_g}{A}$ and V = HA where A is the cross sectional area, then $\phi_g = v_s A$ and thus:

$$\frac{V}{\phi_g} = \frac{AH}{v_s A} = \frac{H}{v_s}$$

Oxygen transfer from an individual gas bubble

The oxygen depletion of the gas phase can be characterised by the characteristic time for oxygen transfer from an individual gas bubble $t_{OT,b}$. In the case of a CSTR this characteristic time is derived from an oxygen balance of the gas phase:

$$\frac{dC_g}{dt} = k_L a \cdot \left(\frac{C_g}{H} - C_l\right) \qquad \text{(E - 2.24a)}$$

in which C_g is the oxygen concentration in the gas phase and C_l is the oxygen concentration in the liquid phase, H_c = Henry coefficient

$$\int \left(\frac{1}{C_g/H - C_l}\right) d\left(\frac{C_g}{H}\right) = \frac{k_L a}{H} \int dt \qquad \text{(E - 2.24b)}$$

so the characteristic time of this differential equation is:

$$t_{OT,b} = \frac{H}{k_L a} \qquad \text{(E - 2.25)}$$

For a bubble column the characteristic time is equal to:

$$t_{OT,b} = \frac{\varepsilon \cdot m}{(k_L a \cdot (1 - \varepsilon))} \qquad \text{(E - 2.26)}$$

in which m is the gas liquid distribution coefficient and ε is the gas hold up.

Heat transfer

The characteristic time for heat transfer t_{HT} follows directly from an overall heat balance over the reactor and in the case of a CSTR or bubble column equals:

$$t_{HT} = \frac{V \cdot \alpha \cdot \rho}{U \cdot A} \qquad \text{(E - 2.27)}$$

in which α the specific heat transfer coefficient, A is the surface of the cooling/heating device and U is the overall heat transfer coefficient.

A problem for the calculation of the heat transfer time will be, however, the accurate estimation of the overall heat transfer coefficient of the reactor. Much literature is known about heat transfer in ungassed stirred tank reactors. For gas-liquid systems, however, the literature is scarce. The effect of stirring and gassing on this parameter is not clear, but values are given ranging from 4000-8000 W m^{-2} K^{-1}.

We have just been through a rather large number of characteristic times and presented you with rather a lot of equations. Some such as $t_{OT} = 1/k_La$ are easy to remember, others are not so easy. It is perhaps more important for you to be able to use these equations rather than remember them. Of course if you can do both then that is a bonus.

Π We suggest you re-read through Section 2.7.2, this time construct yourself a revision sheet of these equations as you read them. We suggest the following format:

Transferred component	**Characteristic time**	**Relation**	**Comment**
Oxygen	t_{OT}	$t_{OT} = 1 / k_La$	where: $k_La = 0.026 \left(\frac{P_g}{V}\right)^{0.4} (v_s)^{0.5}$ (for coalescing systems) $k_La = 0.0016 \left(\frac{P_g}{V}\right)^{0.7} (v_s)^{0.2}$ (for non-coalescing systems)

When you have done that, answer SAQ 2.7 and then move on to the next section where we will examine the characteristic times for conversion.

SAQ 2.7

1) What will happen to the oxygen transfer time from an individual gas bubble if the Henry coefficient doubled its value? Assume k_La remains constant.

2) If the value of k_L remains constant what happens to the bulk oxygen transfer time if the surface are of the bubbles per unit volume (a) is doubled?

3) What is the pumping capacity of a turbine stirrer with a diameter of 0.1 m and a speed of 1 s^{-1}?

4) If the pump described in 3) is fitted into a vessel with a working volume of 1 m^3, what is the circulation time for this system?

5) Using the value of the circulation time you obtained for question 4), calculate the mixing time for the vessel.

6) If the circulation time of a bubble column is 10s and its height is 1m, what is the liquid circulation velocity?

2.7.3 Characteristic times for conversion

Oxygen consumption

The characteristic time for oxygen consumption t_{OC} is derived from the integrated 'Michaelis-Menten' equation, if this type of kinetics is assumed for oxygen consumption:

$$t_{OC} = \frac{K_{O2}}{r_{O2}^{max}} \cdot \ln\left(\frac{C_{O2,0}}{C_{O2,cr}}\right) + \left(\frac{C_{O2,0} - C_{O2,cr}}{r_{O2}^{max}}\right) \qquad \text{(E - 2.28)}$$

in which K_{O2} is the oxygen saturation constant, r_{O2}^{max} the maximal oxygen consumption rate, $C_{O2,0}$ the oxygen concentration at t = 0 and $C_{O2,cr}$ is the critical oxygen concentration. Only the two extremes of this equation will be used:

Zero order type of kinetics if $C_{O2} >> K_{O2}$

$$t_{OC}^{(0)} = C_{O2} / r_{O2}^{(max)} \qquad \text{(E - 2.28a)}$$

First order type of kinetics if $C_{O2} < K_{O2}$

$$t_{OC}^{(1)} = K_{O2}/r_{O2}^{(max)} \qquad \text{(E - 2.28b)}$$

Because in most cases $C_{O2} >> K_{O2}$, the characteristic time for oxygen consumption can be calculated as given by Equation 2.28a. However it is not a *priori* clear what the concentration will be in the reactor. If the characteristic times for oxygen transfer and oxygen consumption are compared (in which for C_{O2}, the saturation concentration C_{O2}^* is taken), it can be shown whether there will be an oxygen limitation or not. If $t_{OC} > t_{OT}$, there is no oxygen limitation and $C_{O2} \approx C_{O2}^*$ in Equation 2.28a is justified. If $t_{OC}^{(0)} < t_{OT}$, there will be an oxygen limitation in the reactor, so the actual concentration in the reactor will be very low (approximately equal to K_{O2}). For comparison of the oxygen consumption time with other characteristic times, for example the circulation time, $t_{OC}^{(1)}$ has to be used.

Substrate utilisation

Basically, for this characteristic time the same assumptions can be made as for the oxygen consumption time. In batch processes, the substrate concentration C_s can be taken to be equal to the substrate concentration at inoculation time. For continuous or fed-batch processes, the actual concentration in the reactor has to be taken into account for the calculation of the substrate utilisation time. In that case often 'Michaelis-Menten' kinetics will be used. Because in most systems under investigation $C_s >> K_s$, the characteristic time for substrate utilisation t_{sc} reads:

$$t_{sc} = \frac{C_{s,0}}{r_s^{max}} \qquad \text{(E - 2.29)}$$

in which r_s^{max} is the maximal substrate consumption rate and $C_{s,0}$ is the substrate concentration at $t = 0$.

Growth

In batch processes, the characteristic time for growth t_G can be obtained from the reciprocal specific growth rate:

$$t_G = (\mu_{max})^{-1} \qquad \text{(E - 2.30)}$$

in which μ_{max} is the maximal specific growth rate. Note that for continuous processes the specific growth rate (μ) = the dilution rate (D). Thus for continuous processes, this characteristic time becomes equal to the reciprocal dilution rate.

Heat production

To calculate the characteristic time for heat production t_{hP}, a temperature difference has to be used to define a zero-order heat production. If the temperature difference between the reactor contents and the cooling/heating liquid is used it follows that if $t_{ht} = t_{hP}$, all the heat produced is removed from the system (where t_{ht} = characteristic time for heat transport). Then, the heat production time can be calculated from:

$$t_{hP} = \frac{\rho \,.\, \alpha \,.\, \Delta T}{r_{hM} + r_{hR}} \qquad \text{(E - 2.31)}$$

in which ΔT is the temperature difference between reactor interior and cooling/heating liquid, r_{hM} is the heat production by the micro-organisms [$r_{hM} = 460 \,.\, 10^3 \,.\, r_{02}^{max} \, V$ (J)], and r_{hR} is the heat production by the CSTR or bubble column. In the case of a CSTR the heat production is approximately equal to the gassed power input P_g. ρ = density of liquid, α = specific heat transfer coefficient.

Π Again it would be useful to re-read this section and produce for yourself a revision sheet summarising the relationships giving the characteristic times for conversions.

SAQ 2.8

1) If the concentration of oxygen is x mol m^{-3} and an organism can consume this oxygen maximally at a rate of $\frac{x}{40}$ mol m^{-3} s^{-1}, what is the characteristic time for oxygen consumption. Note that the saturation constant for oxygen for this organism is $\frac{x}{1000}$ mol m^{-3}.

2) The characteristic time for oxygen consumption in a CSTR system has been calculated to be 10s. If the characteristic time for oxygen transfer into the system is 15s, will there be an oxygen limitation in the reactor?

3) If the k_La value of the system described in 2) is doubled, will there be oxygen limitation in the reactor?

4) The flow rate of media into a continuous process is F m^3s^{-1}. If the volume of the vessel is Vm^3 and F $= \frac{V}{5000}$ m^3 s^{-1} , what is the characteristic time for growth in the vessel?

Summary and objectives

This has been a long chapter in which we have examined the strategies that can be applied to scale up. We particularly focused on three types of approach. One which employed 'rules of thumb' involved a variety of parameters as the criterion used for scale up. Popular in this approach is to use a constant P/V ratio as the scale up criterion, although other parameters (eg k_La, mixing time, stirrer speeds, shear forces etc) may be used. We also examined the application of dimensional analysis to scale-up. In this approach analysis of the dimensions of the parameters which influence a process enables them to be grouped together into dimensionless numbers (or coefficients). It is these dimensionless numbers which are used as the criterion for scale up. In the final part of the chapter we examined what has been referred to as a regime analysis. Typical of this approach is to describe a process in terms of characteristic times. These times may then be used as scale up criteria or provide evidence for rate limitations.

Now that you have completed this chapter you should be able to:

- identify from supplied data if the relationship between selected criterion (such as constant P/V) and process performance (such as product yield) is dependent or independent of scale of operation;
- calculate k_La values from supplied data for a variety of conditions including coalescing media and non-coalescing media in CSTRs or in bubble columns;
- apply Buckingham II theorem to determine the minimum number of dimensionless numbers from a supplied set of parameters;
- use Rayleigh's method for determining dimensionless numbers from a supplied set of parameters;
- identify a variety of dimensionless numbers;
- apply similarity principles to determine parameter values from supplied data;
- calculate a variety of characteristic times from supplied data;
- use terms relating to characteristic times appropriately.

Scale down and its application to reactor design

Scale down and its application to reactor design

3.1 Introduction

In the previous chapter we examined several approaches to the problems encountered in scaling-up a process from laboratory scale to production scale. In these approaches, the usual sequence is to establish the performance of the system on a laboratory scale and then try to use the conditions established in the laboratory scale to design the production scale process. It is perhaps obvious that this is the natural sequence of events since most processes begin development on a laboratory scale and most of the early analysis is done on this scale. Intellectually, however, this is not the only sequence (ie laboratory → pilot → production scale) we have to follow. We can, in principle, establish the conditions the micro-organisms meets at the production scale and then try to simulate these conditions at the laboratory scale in order to conduct experiments. Such an approach is referred to as scale down. Thus the scale down approach is one in which the conditions met at the production scale are simulated at a laboratory scale and this laboratory scale operation is used to study the effects of changing parameters on performance. The ultimate aim of the scale down approach is to improve performance of the production process.

structure of this chapter

This chapter examines the scale down approach to scale up issues. First, the basic strategy of scale down is examined followed by a more detailed discussion of scale down based on regime and process analysis. In the final part of the chapter we provide some examples of the application of the scale down approach including its application to gluconic acid production, microbial desulphurisation of coal, butanol and xanthan gum production and yeast culture. In contrast to Chapter 2, this chapter is quite brief and you should be able to complete it in one sitting.

3.2 The basic strategy of scale down

From the description given above, it should be self evident that we need good rules to scale down phenomena like mixing, gas bubble residence time, mass transfer fluid shear etc. It is not our intention here to go through all these rules since effectively they are analogous to those we have covered in Chapter 2. For example, we gave relationships for a large series of characteristic times and we described relationships between k_La, the power consumption, the volume of the vessel and superficial gas velocity and so on. Here we which to establish an over-all strategy for scale down.

A first step to developing a strategy for scale down was taken in 1966 when it was suggested to split up the fermentation process into individual steps and to establish the effects of these separate steps on the microbial process. For example Oldshue (1978 Process Biochem 12, 16) suggested that for penicillin fermentation, the following elements are essential:

- effectiveness of the mixer;
- influence of the oxygen transfer rate on microbial growth and product formation;
- the effect of the cell concentration on the mass transfer rate (change of viscosity during fermentation);
- the effect of the fluid shear upon microbial growth.

A more theoretical approach was found to be, to simulate the production scale by a mathematical model. A block diagram representing such a model is represented in Figure 3.1. In this figure we have indicated the components that need to be considered relating to oxygen availability within the fermentor. You will notice many of the terms you might expect such as hold up, bubble age distribution, mass transfer, mass balance etc. We need not however concern ourselves with the mathematical details of this approach. What is important is that in this model the evaluation of the production scale kinetic behaviour of the micro-organism is an essential step, so one has to know the influence of the production scale conditions upon the growth and product formation of the micro-organism. This requires the design of reliable scale down experiments to establish these influences. However, kinetic models describing the behaviour of micro-organism under different and transient conditions are scarce. Development of these models and determination of their parameters is very time consuming. Therefore, this method is not yet of practical use for industry.

limited knowledge of kinetic models in transient conditions

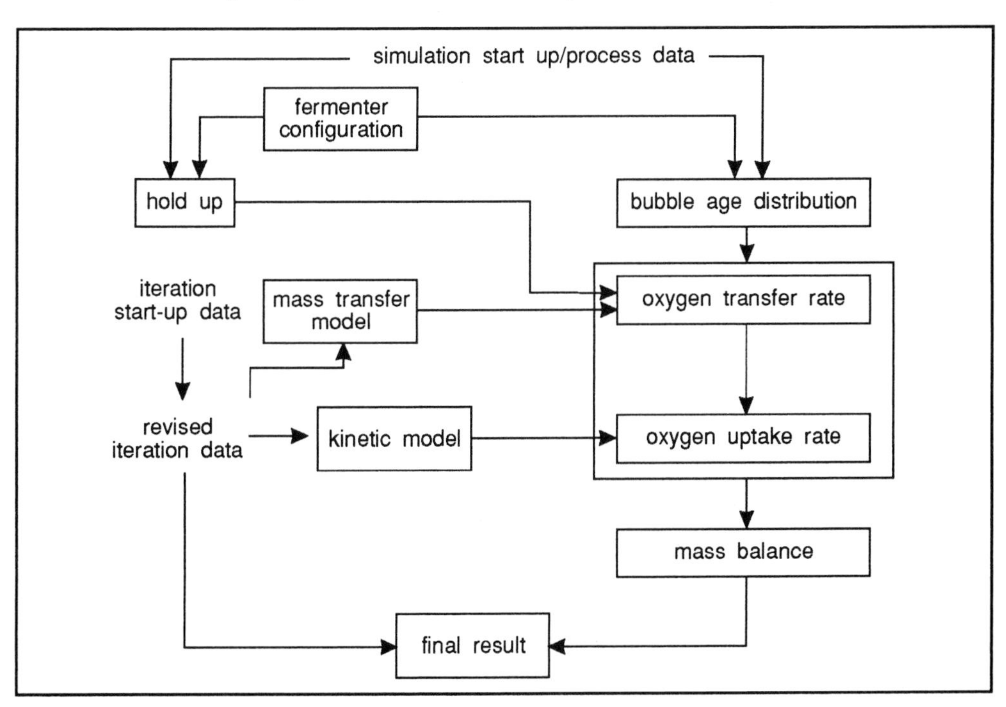

Figure 3.1 Block diagram for fermentation design by a simulation method (redrawn from Ovaskainen Lundell and Laiho 1976, Proc Biochem 11, 37-39).

To carry out scale down experiments, it may not be necessary to scale down the complete process. Furthermore, scale down maybe combined with regime analysis. With the help of regime analysis (determination of the time constants and therefore, the rate limiting steps) it is possible to predict the rate limiting steps of a process at the industrial scale. In the next section scale down based on regime analysis will be discussed in more detail.

3.3 Scale down based on regime analysis

In this procedure, introduced by Oosterhuis in 1984, four steps can be distinguished (see also Figure 3.2).

- regime analysis of the process at production scale;
- simulation of the rate limiting mechanisms at laboratory scale;
- optimisation and modelling of the process at a laboratory scale;
- optimisation of the process at production scale by translation of the laboratory conditions.

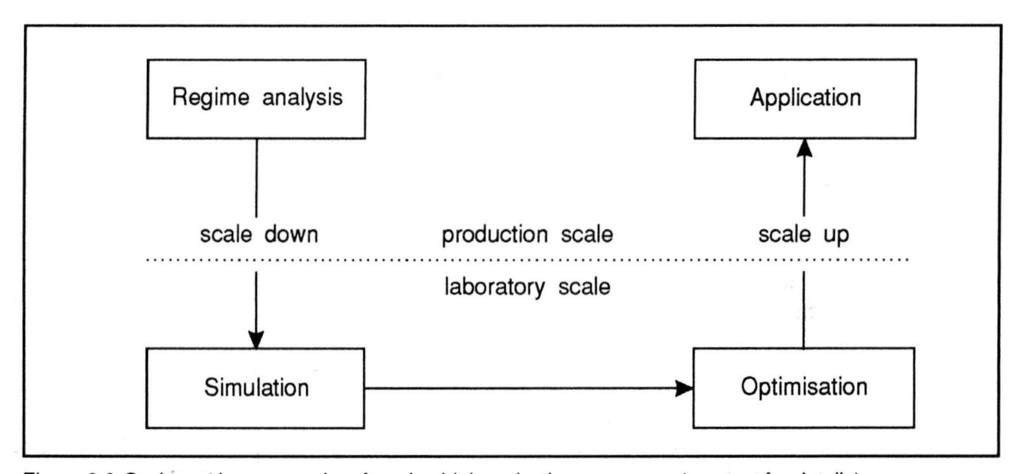

Figure 3.2 Scale-up/down procedure for microbial production processes (see text for details).

question to be answered by regime analysis

Thus the regime analysis must give answers to several questions. What mechanism of the process is rate limiting, in other words what is the ruling regime? Is the regime ruled by one mechanism (pure regime) or more mechanisms (mixed regime)? Will there be a change in regime? The analysis has to allow for changes of scale, changes in process parameters and the course of the process. An important factor in the performance of the micro-organisms is whether the existing regime will be characterised by nutrient limitation or fluctuations in the environment of the cells. The yield of biomass or products can, for example, be affected by fluctuations in the concentration of a component, in the pH, in temperature or shear rate.

From the results of the regime analysis it can be concluded what mechanisms or features require further investigation on a small scale. The most important criterion for the experiments on a laboratory scale is that they have to be representative of the conditions at production scale. This determines the limits of small scale experiments.

Let us test whether you have understood this phase of the strategy of scale down based on regime analysis by attempting SAQ 3.1. You will need to use some of the relationships we described in Chapter 2.

SAQ 3.1

You are seeking to use a scale down approach to optimise a continuous production process. To do this you have conducted a regime analysis on the production scale process. Part of the data you have obtained is presented below. From this data determine what mechanisms of the process is rate limiting and suggest what experiments might be conducted on a laboratory scale.

$k_La^{20} = 0.04\ s^{-1}$

$r_{O2}^{max} = 0.1\%$ of $C_{O2}^{*}\ s^{-1}$

$C_{O2} = 10\%$ of C_{O2}^{*} (where C_{O2}^{*} = saturation concentration of O_2)

$K_{O2} = 0.01\%$ of C_{O2}^{*} (where K_{O2} is the oxygen saturation constant of the organism being used)

$C_{s,0} = 1\ mol\ m^{-3}$ (where $C_{s,0}$ = input substrate concentration)

$K_s = 10\ mmol\ m^{-3}$ (where K_s = substrate saturation constant)

$r_s^{max} = 0.001\ mol\ m^{-3}\ s^{-1}$ (where r_s^{max} = maximum substrate consumption)

$D = 1\ h^{-1}$ = dilution rate

In the response to SAQ 3.1, we made comments about the optimisation process.

Optimisation of the process on a laboratory scale and modelling of the investigated features is the third step of the scale down procedure. In optimisation one has to keep in mind that the optimised situation has to be translated to the production scale. Consequently, not all results of the optimisation can be used (eg when an extremely large stirrer speed is required).

With regard to the effect of fluctuations, the following remarks can be made. Fluctuations tend to increase during scale up (decreased mixing), resulting in transient conditions for the micro-organisms. Much is known about modelling of balanced growth and product formation. However, little is known about growth and product formation under transient conditions. The subject has however been reviewed by Barford, Pamment and Hall (1982 'Lag Phases and Transients' ed Bazin 'Microbial Population Dynamics' CRC Press, Florida 55-89) if you wish to pursue the topic further. We do not intend to examine this aspect in detail here except to say that a distinction must be made between empirical and mechanistic models. Most of the models described in the literature are empirical, do not predict and add little to the fundamental understanding of microbial dynamics. Few mechanistic models have been presented in the literature. This is due to the fact that the biochemical and physiological information, which forms the basis of the mechanistic models is insufficient. Therefore, experimental procedures are still very important.

In the fourth step the optimised laboratory conditions are translated to production scale. Models formed in the previous step can be used for this purpose. Rules used to scale down the process can now be used in scaling up the experimental conditions. The success of the scale up depends on the success in designing representative scale down experiments.

3.4 Scale down based on process analysis

In principle, the scale down step, ie the step from regime analysis to experimental simulation, is based on constant characteristic times. As such, it is only a variant of the scale down, based on constant dimensionless numbers.

For biotechnological processes the behaviour of microbial cultures under large scale (transient) conditions is largely unknown. This makes it impossible to express all mechanisms in dimensionless numbers, whereas it is often possible to make an estimation of, for example, the characteristic time of mechanisms.

reasons why regime analysis is more applicable than dimensional analysis

We may conclude however that if the behaviour of the microbial culture is not the bottleneck of the process, an analysis of engineering problems may be based on dimensionless numbers. It must be remarked that in engineering problems a complete description by means of dimensionless numbers may also be impossible. It can only be used in scale down when used in combination with mechanistic analysis, the similarity principle and regime analysis. This results in an extension of the scheme presented in Figure 3.2. Therefore, this approach is better described as process analysis. Figure 3.3 shows the interactions of the mechanisms used in the analysis of the process. Complementary knowledge may be supplied from rules of thumb, literature data and correlations and experience. The process analysis can be regarded as a systematic method, not restricted to regime analysis alone, to analyse the performance of a large scale process, resulting in an experimental design for small scale simulation of large scale conditions.

The message we are trying to convey in this discussion and by Figure 3.3 is that a single strategy for designing a reactor (or any other stage of a bioprocess) is unlikely to be able to solve all of the problems. We need to use all of the strategies available if we are to be successful. A key feature of process analysis is therefore to use a variety of strategies and to call upon published data and correlations and to use experience as much as possible.

We would however like to emphasise the differences in dimensional analysis.

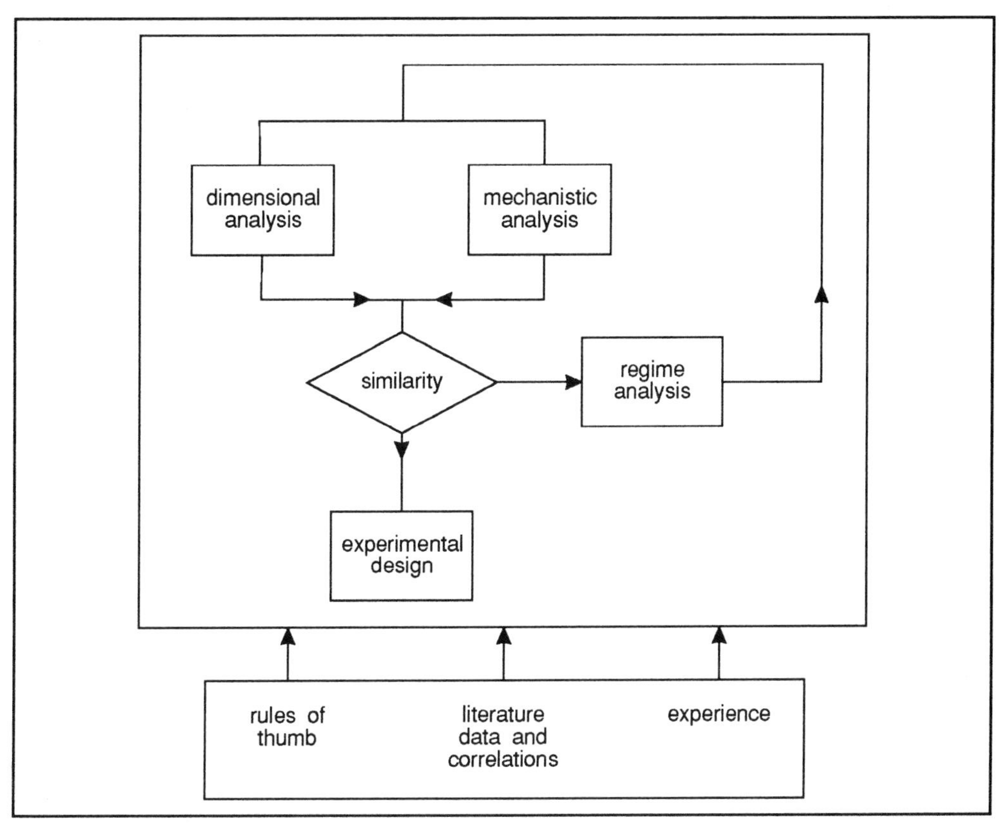

Figure 3.3 Interaction of the methods used in process analysis (see text for discussion).

Π Write down a fundamental difference between dimensional analysis and mechanistic analysis and compare your response with our discussion below.

Whereas dimensional analysis starts from a list of relevant parameters, the mechanistic analysis is based on a list of mechanisms involved. The rate of these mechanisms has to be expressed in characteristic parameters, like fluxes, pressures, heights or times. Further analysis may be based either on the characteristic parameters or on the ratio of these parameters, ie dimensionless numbers. To obtain the same behaviour of systems at different scales the systems have to be similar. As mentioned in Chapter 2, four similarity states are important in chemical engineering: geometrical, mechanical, thermal and chemical, each state necessitating the previous ones. However, in engineering problems it is, on the whole, impossible to satisfy the similarity demand, resulting in the fact that neither all the characteristic parameters, nor all the dimensionless numbers can be kept constant during scale up or scale down. Therefore, it has to be decided which mechanisms are the most important in the system under investigation: in other words what is the ruling regime. So, in fact we, again, end up with scale down based on regime analysis.

3.5 Application of scale down methodology

3.5.1 Sources of information

In this section we are going to describe some examples of the application of scale down methodology. Since these are based on published examples we have cited the key references in Table 3.1. We do not propose that you follow up all of these original publications but by providing the references, you have the opportunity to do so if you so wish.

The method of scale down based on regime analysis has been applied to a number of microbial processes. It was introduced by Oosterhuis (1984) in a study on the optimisation of the gluconic acid production. Huber *et al* (1983) and Bos *et al* (1985) used regime analysis to design an installation for the microbial desulphurisation of coal. Schoutens (1986) used the scale down approach to design a large scale process for the production of butanol from whey permeate. Oosterhuis *et al* (1985, 1987) and Olivier and Oosterhuis (1988) designed a new bioreactor for the fermentation of xanthan gum. Sweere (1988) applied the same principles for determination of the environmental effects which will be present in a large scale bakers' yeast production in a bubble column.

In all of these examples the scale down approach has contributed to optimising the microbial process or reactor configuration. It is possible, using this methodology, to obtain solutions to the complex problems that usually characterise the scale up of many biotechnological processes as will be shown in the following sections.

Bos P, Huber T F, Kos C H, Ras C and Kuenen J G (1985)	Int Symp on Biohydrometallurgy, Vancouver
Huber T F, Kossen, N W F, Bos P and Kuenen J G (1983)	Proc Symp Biotechnol Res in The Netherlands, Nov 22, Delft
Olivier A P C and Oosterhuis N M G (1988)	8th Int Biotechnol Symp, Paris
Oosterhuis N M G (1984)	PhD Thesis, TU Delft, The Netherlands
Oosterhuis N M G and Koerts K (1985)	Eur Patent Appl 0185407
Oosterhuis N M G and Meyer P D (1987)	Proc 4th Eur Conf on Biotechnology 240
Schoutens G M (1986)	PhD Thesis, TU Delft, The Netherlands
Sweere A P J (1988)	PhD Thesis, TU Delft, The Netherlands

Table 3.1 Literature sources of examples of the application of scale-down based on regime analysis.

3.5.2 Gluconic acid production

reactor used for gluconic acid production

The process of gluconic acid production by *Gluconobacter oxydans* was studied by Oosterhuis (1984). This aerobic process was already running at 25 m^3 scale in traditional stirred tank reactors, equipped with two standard Rushton type impellers (see Figure 3.4).

Table 3.2 presents the relative dimensions of the production-scale fermentor. Characterisation of the process by regime analysis, based on estimations, gave the results as presented in Table 3.3.

Let us examine these tables in a little more detail. Table 3.2 will give you some idea of the dimensions of vessels used in real processes. Of course you should anticipate differences between different processes but the dimensions given in Table 3.2 are fairly typical. Thus quite often the impeller has a diameter of about a third of the vessel and typical gas flow rates are of the order of 0.5 $vv^{-1}min^{-1}$. Read through the table carefully and note the dimensions and ratios given.

Table 3.3 gives the characteristic times of a variety of the mechanisms which are important in gluconic acid production.

Π From the data presented in Table 3.3, is oxygen availability likely to be rate limiting and are oxygen gradients likely to be established in the system?

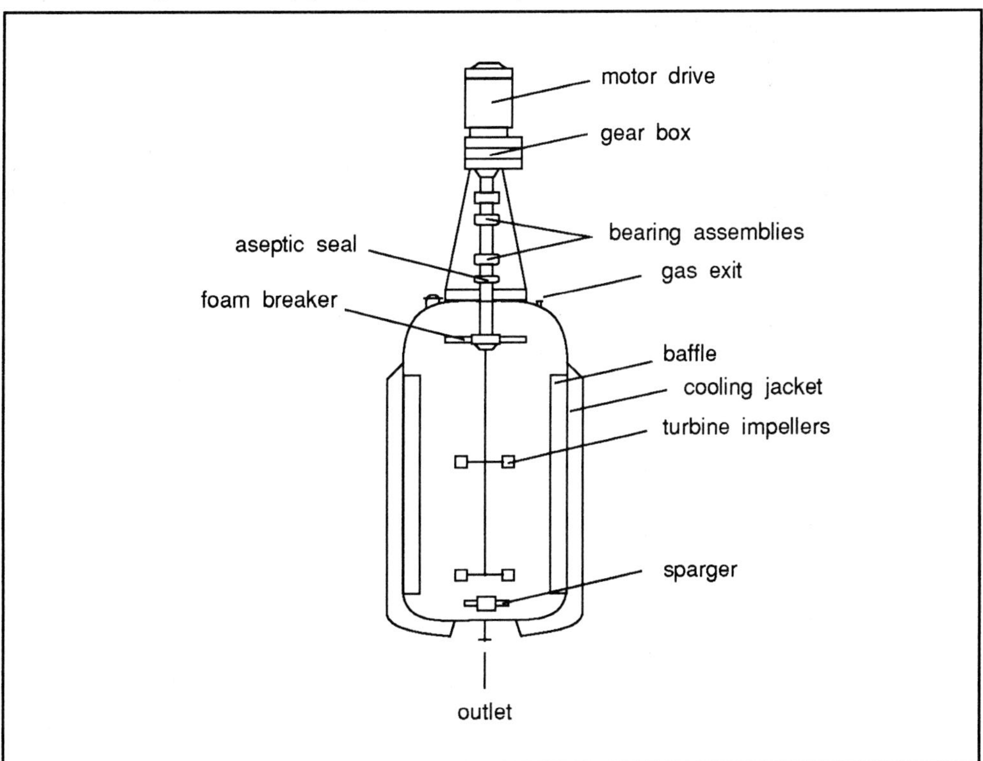

Figure 3.4 Stirred tank reactors used in gluconic acid production. (Not to scale, note that the tank volume is usually very much larger in proportion to the motor drive and bearing assembles).

Impeller/vessel diameter	0.32
Number of impellers	2
Impeller blade width/impeller diameter	0.2
Impeller speed	1.3 or 2.6 s^{-1}
Baffle diameter/vessel diameter	0.09
Liquid height/vessel diameter	up to 1.8
Gas flow/reactor volume * time	up to 0.5 vvm
Liquid volume	up to 25 m^3

Table 3.2 Dimensions of the production scale reactor for gluconic acid production.

Transport phenomena	**characteristic time(s)**
Oxygen transfer	5.5 (non-coal) -11.2 (coal)
Circulation of the liquid	12.3
Gas residence	20.6
Transfer of oxygen from a gas bubble	290 (non-coal) -593 (coal)
Heat transfer	330 -650
Conversion phenomena	
Oxygen consumption	
zero order	16
first order	0.7
Substrate consumption	$5.5 * 10^4$
Growth	$1.2 * 10^4$
Heat production	350

Table 3.3 Characteristic times (in seconds) of the mechanisms which are important in the gluconic acid production (non-coal = non-coalescing system; coal = coalescing system).

The following conclusions for the production at large scale can be drawn from Table 3.3. The times for oxygen consumption (16 s) and oxygen transfer (11.2 s) to the liquid phase are of the same order of magnitude. Therefore, oxygen limitations may occur. Also the liquid circulation time is of the same order of magnitude, and thus oxygen gradients are likely to occur.

Π From the data relating to gas residence time and the characteristic time for oxygen transfer from a gas bubble, will the gas in the bubbles become exhausted of oxygen?

From a comparison of the gas residence time and the time for oxygen transfer from the gas phase it is clear that no exhaustion of the gas phase will occur.

Π Use the data in Table 3.3 to decide if heat production and heat transfer balance each other and whether or not temperature gradients will be established.

Since heat transfer (characteristic time 330-650 s) and heat production (characteristic time 350 s) are similar, heat transfer will balance the heat production. If the time constants for heat production and heat transfer are compared with the liquid circulation time it can be concluded that temperature gradient will not be present in the fermenter. It has to be remembered that the correlations used only give a rough estimation of the characteristic times, so only the order of magnitude can be considered and compared.

This regime analysis formed the basis of small-scale investigations in which the effects of oxygen gradients on the microbial conversion process were studied. Based on these experiments it was possible to optimise the (production scale) conditions with respect to oxygen transfer and liquid circulation. However, in this case, realisation at production scale of such optimisation was restricted due to the influence of the oxygen transfer (consumption) on the heat transfer (production). From this example it can be concluded that small scale experiments can avoid a lot of work which in the end will not bring the desired results. Further optimisation of this process asks for drastic adaptations of the full scale reactor configuration or even for complete new reactor design.

3.5.3 Design of an installation for the microbial desulphurisation of coal

Huber *et al* (1983) and Bos *et al* (1985) used regime analysis based on characteristic times to design an installation for the microbial desulphurisation of coal. For optimal design of the reactor two major conditions must be met. 1) Biomass is the catalyst for the oxidation of pyrite to sulphate, so biomass limitation must be prevented. 2) The environment of the micro-organisms must be optimal for pyrite oxidation.

In relation to the first point, Huber *et al* (1983) stated that a reactor configuration consisting of a mixed flow reactor followed by a plug flow reactor would be the most adequate, because the pyrite oxidation obeys first order kinetics (Levenspiel, 1972). A plug flow reactor alone would result in wash out of biomass. Therefore, an intensively mixed fermentor has to be used first in order to generate an effective inoculum. Limitation of biomass can then be avoided if the residence time in this well stirred fermenter is greater than the characteristic time of biomass growth:

$$\tau > t_G \qquad \text{(E - 3.1)}$$

Optimal conditions for pyrite oxidation demanded that depletion of oxygen and settling of the coal particles had to be prevented throughout the fermenter. To achieve this, three conditions had to be met:

- no overall oxygen depletion:

$$t_{OT,l} < t_{OC} \qquad \text{(E - 3.2)}$$

where $t_{OT,L}$ is the characteristic time of transfer of oxygen into the liquid, t_{OC} is the characteristic time for oxygen consumption

- no gradients in oxygen concentration:

$$t_{m,l} < t_{OT,l} \qquad \text{(E - 3.3)}$$

$t_{m,l}$ is the characteristic mixing time of the liquid

- no sedimentation of the coal:

$$t_{m,l} < t_{sett} \quad \text{(E - 3.4)}$$

t_{sett} is the characteristic settling time

If the kinetics of the desulphurisation reaction are known the oxygen demand can be calculated (Huber *et al* 1983). Based on the required capacity of the installation and the conditions in Equations 3.1 to 3.4, the reactor can be designed. Table 3.4 gives the values of the characteristic times of the process.

Parameter	Characteristic time(s)
Liquid mixing	60-118
Oxygen transfer	80-641
Oxygen consumption	$3.4 * 10^5$
Settling of the particles	$2.5 * 10^4$
Growth	$8.6 * 10^4$

Table 3.4 Characteristic time(s) relevant to the microbial desulphurisation of coal.

SAQ 3.2

From the data presented in Table 3.4 will:

1) oxygen be a limiting factor;

2) the settling of the coal particles be a problem;

3) oxygen gradients be a problem?

The regime analysis described above led to the design of an installation with the specifications given in Table 3.5. Again we have included this data to give you some idea of the scale of a real process.

Configuration	cascade of 10 pachuka tank reactors
Capacity	100 000 ton year^{-1}
Gas velocity	0.02 - 0.003 m s^{-1}
Diameter of the pachuka tank reactor	10 m
Height of the pachuka tank reactor	20 m
Pyrite content	0.5%
Pyrite removal	90%

Table 3.5 Specification of the installation for the microbial desulphurisation of coal.

3.5.4 Design of a butanol production process

airlift loop and fluidised bed reactors

Schoutens (1986) studied the design of a large scale process for the production of butanol from whey permeate with immobilised *Clostridium* cells. For the design of scale models of an airlift loop reactor (GLR) and a fluidised bed reactor (FBR) with liquid circulation, regime analysis of a FBR reactor of 50 m^3 and a GLR of 65 m^3 was carried out. The characteristic times of the liquid mixing, the liquid circulation and the product formation are given in Table 3.6. Note that the features of a laboratory scale gas lift loop reactor are also included in this table.

Parameter	FBR	GLR 1	GLR 2
Volume (m^3)	50	65	0.010
Bead fraction (-)	0.45	0.35	0.35
Dilution rate (h^{-1})	0.3 - 0.5	0.2 - 0.4	0.2 - 0.4
Superficial liquid velocity ($m\ s^{-1}$)	$5\ 10^{-3}$	0.5 - 1.0	0.2 - 0.3
Superficial gas velocity ($m\ s^{-1}$)	-	$(1 - 3)\ 10^{-2}$	$(1 - 3)\ 10^{-2}$
Production time(s)	$(7.2 - 12)\ 10^3$	$(9 - 18)\ 10^3$	$(9 - 18)\ 10^3$
Liquid circulation time(s)	600-1200	20-40	5-20
Axial dispersion time(s)	$(6.9 - 69)\ 10^5$	700-1000	60-500

Table 3.6 Specifications and characteristic times of a fluidised bed reactor (FBR) with recycle and two gas lift loop reactor (GLR) for the continuous butanol fermentation. GLR 1 = large scale; GLR 2 = laboratory scale.

Π From the data presented in Table 3.6, can it be concluded that the laboratory scale reactor adequately represent the production scale process. (Hint, does liquid circulation times etc change substantially in the scaled down process)?

From the characteristic times of the 65 m^3 GLR it can be concluded that a change of the hydrodynamical regime is unlikely when scaling down to laboratory scale. However you should note that one criterion that determines the design of the model GLR is to avoid wall effects of the bubbles and the beads. In practice the laboratory scale GLR was chosen with an external loop and a volume of 15 litres and where the height was 1 m and the diameter of the riser and downcomer were 0.08 m.

3.5.5 Xanthan gum production

effects of changing viscosity

In order to scale up the xanthan production by *Xanthomonas campestris* which is characterised by oxygen consumption in an extremely viscous broth, a method to scale up this process has been suggested which involve experiments at laboratory and pilot plant scale, which should have the same behaviour as a full-scale plant. This was in order to meet the same environmental conditions at the various scales. Although such an approach has its value, a more mechanistic approach was followed by Oosterhuis *et al* (1985, 1987) and Olivier and Oosterhuis (1988). Using the principles of scale down, the rate limiting steps for the xanthan gum fermentation were determined. It could be concluded that during the fermentation in a classical stirred tank reactor the characteristic time for oxygen transfer changes from 15 (s) when viscosity is low to 280

(s) when viscosity is high. Due to the build-up of viscosity by product formation, a constant change of regime takes place during the (batch) fermentation. This is not restricted to the production scale, but occurs also at the laboratory scale. It was therefore concluded that a stirred tank reactor in the 'traditional' configuration, even at laboratory scale, is not suitable to carry out such processes.

It is evident that another approach has to be followed to tackle the problems in this type of fermentation. This is because adaptation of the process to the reactor or even slight adaptations of the reactor do not give the desired results. An alternative is to design a complete new reactor, which meets the production requirements. In this case the new type of reactor was the Pumped Tower Loop Reactor, illustrated in Figure 3.5.

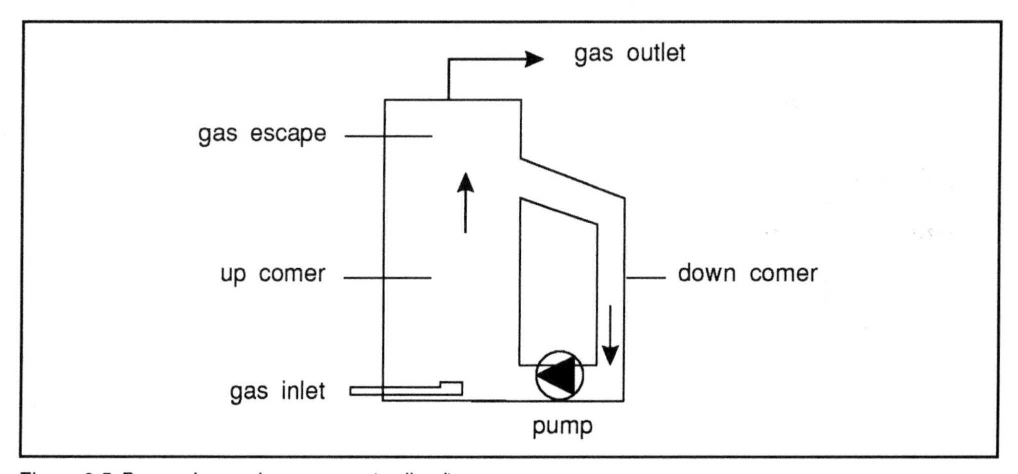

Figure 3.5 Pumped tower loop reactor (stylised).

Essentially a Pumped Tower Loop Reactor (PTLR) is an airlift reactor in which the airlift principle is assisted by a pump. Thus in xanthan production, because of the increase in viscosity, the driving force of the gas on which the airlift principle depends, is not usually sufficient to keep the viscous fluid in motion. Furthermore because of the viscosity there is a stratification of gas in the middle of the column of the airlift reactor leading to 'canals' while fluid in the vicinity of the walls remains stagnant. In the PTLR, the liquid is kept in constant motion by using a specially designed pump placed at the connection between the riser and downcomer.

The performance of this PTLR was tested at 30 litre and 4000 litre scale. It was concluded that oxygen transfer under the defined flow conditions in this reactor is much more efficient than in a stirred tank reactor (Oosterhuis and Meyer, 1987). In addition, scale up from 30 litres to 4000 litres was possible without change of regime and further scale up can be simulated due to good modelling possibilities of such a reactor. Figure 3.6 shows results in the 4000 litre reactor compared with those in a 1500 litre stirred tank reactor and with laboratory data from stirred tank fermenters. From these results it can be concluded that the performance of the PTLR in the production of xanthum gum is much better than that of the CSTR.

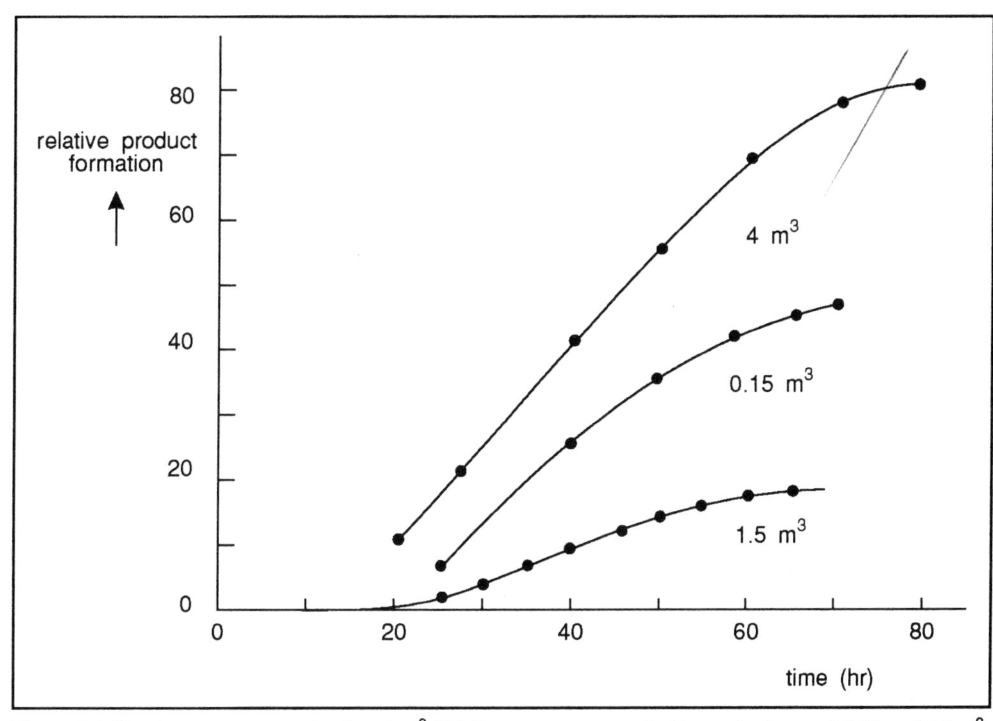

Figure 3.6 Xanthan gum production in a 4 m^3 PTLR reactor compared with results from a 0.015 and 1.5 m^3 stirred tank fermenter (data from Olivier and Oosterhuis 1988 8th Int Biotechnol Symp, Paris).

3.5.6 Bakers yeast production

yeast metabolism of glucose depends on oxygen/glucose availability

The metabolism of sugars by bakers yeast is sensitive to changes in glucose concentration and changes in oxygen concentration. A high glucose concentration or a low oxygen concentration will result in the production of ethanol by the yeast. At high oxygen levels, CO_2 is produced. It is important to know whether limitations or gradients of oxygen or substrate will occur. Regime analysis was carried out by Sweere (1988) for a bubble column fermenter of 150 m^3. Tables 3.7 and 3.8 give the characteristic times of the mechanisms which may play a role in the performance of the fed batch fermentation. A discrimination was made between mechanisms which have a clear effect on the process (Table 3.7) and mechanisms which do not (Table 3.8).

Mechanism	Definition	Magnitude(s)
Mixing:		
Liquid phase	$t_{m,l} = L^2/D_{E,l}$	$10^1 - 10^3$
Gas phase	$t_{m,g} = L^2/D_{E,g}$	$10^1 - 10^3$
Liquid circulation	$t_{cir} = 2L/v_{cir}$	$10^1 - 10^2$
Gas flow	$t_g = (1-\varepsilon)L/v_s$	$1 - 10^2$
Oxygen transfer:		
Liquid phase	$t_{OT,l} = 1/k_La$	$1 - 10^2$
Gas phase	$t_{OT,g} = \varepsilon m/(k_La.(1-\varepsilon))$	$10^3 - 10^6$
Substrate consumption	$t_{SC} = C_s/r_s$	$10^1 - 10^2$
Oxygen consumption	$t_{OC} = C_{o,L}/r_o$	1
Substrate addition	$t_{sa} = V.C_s/(F_s.C_{s0})$	$10^1 - 10^2$

Table 3.7 Characteristic times of the relevant mechanisms in bakers yeast production in a bubble column fermenter.

Π Why do we claim that the mechanisms listed in Table 3.7 are relevant to bakers yeast production while those in Table 3.8 not?

Mechanism	Definition	Magnitude(s)
Fed batch process	t_p	10^5
Growth of biomass	$t_x = 1/\mu$	10^4
Heat transfer	$t_{ht} = V.\rho_L.c_p/(U.A.\Delta T)$	10^3
Heat production	$t_{hp} = V.\rho_L.c_p/(r_h + P/V)$	10^3
Micro-mixing:		
Diffusion	t_D	$10^{-2} - 1$
Turbulent erosion	t_{te}	$1 - 10^2$
Laminar stretching	t_{ls}	10^{-3}

Table 3.8 Characteristic times of the bakers yeast production which are not relevant for the final performance of the reactor.

You should have put forward two types of reasons. Firstly we explained that oxygen and substrate concentrations greatly influence the metabolism of yeast. At low substrate/high oxygen tensions the substrate (glucose) is oxidised to CO_2 and H_2O whilst at high substrate/low oxygen glucose is converted to ethanol. Thus it is not surprising to find mechanisms involved in influencing oxygen and substrate availability listed in Table 3.7.

The second set of reasons relate to the characteristic times of these mechanisms. Many are very similar, thus it is very important to be able to measure/calculate these accurately if we are to ensure a particular substrate/oxygen regime. Obviously if $t_{OT,l}$ is similar to t_{oc} then it is critical that we can predict these accurately since this will govern whether CO_2 or ethanol is produced by the yeast. Likewise if $t_{m,l}$ is similar to $t_{OT,l}$ then we can predict that oxygen gradients will occur and this too will influence whether CO_2

or ethanol production takes place. For this reason it is important to examine these characteristic times very critically and examine the influence of a wide variety of parameters on these characteristic times. We will examine this principle by examining the effects of superficial gas flow rate, height to diameter ratios and substrate concentration in various bubble column reactors.

The characteristic times of the liquid mixing and circulation, mass transfer and oxygen consumption are shown as functions of the superficial gas flow rate in a bubble column in Figure 3.7. These graphs have been calculated for a fed-batch production at a volume of 120 m^3 and a height-to diameter ratio of 4. They allow easy comparison of the rates of the various mechanisms.

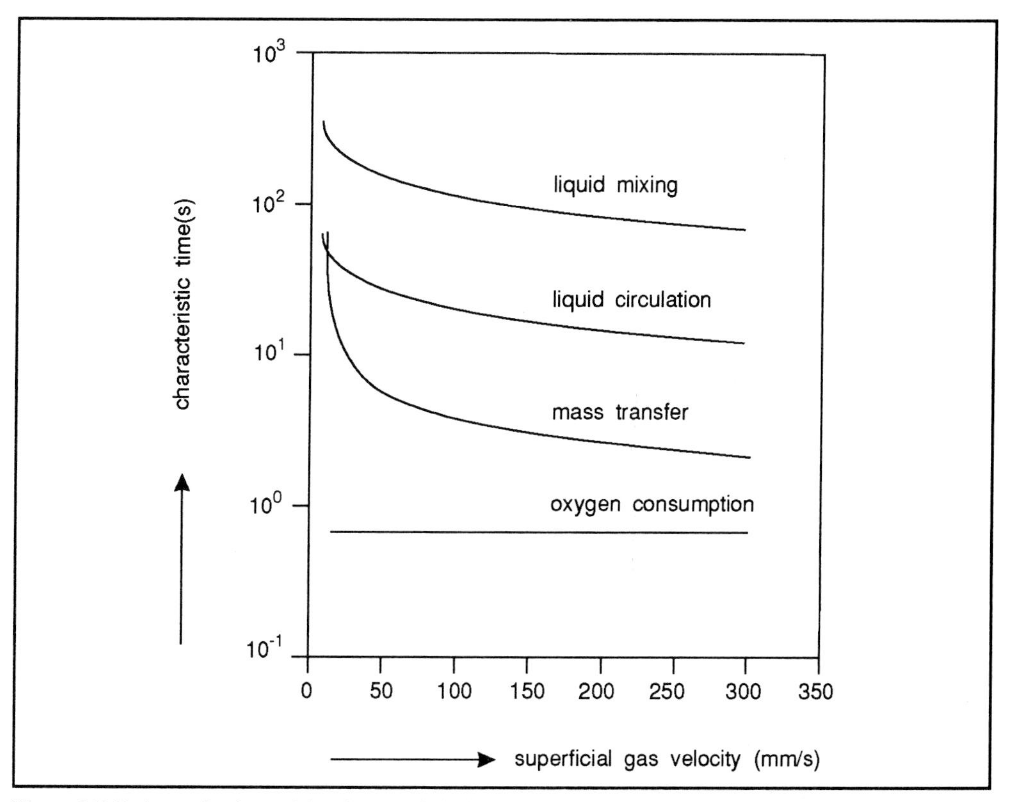

Figure 3.7 Estimated values of the characteristic times of fed-batch bakers' yeast production as a function of the superficial gas velocity in a 120 m^3 bubble column reactor with a height-to diameter ratio of 4 (see text for details). Data from Sweere 1988, PhD Thesis, TU Delft, The Netherlands.

SAQ 3.3

From the data presented in Figure 3.7 will:

1) oxygen limitation occur in the vessel?

2) gradients of oxygen be established in the vessel?

3) ethanol or CO_2 likely to be the major product of substrate metabolism?

The effect of the height-to-diameter ratio, H/D, on the characteristic times is shown in Figure 3.8.

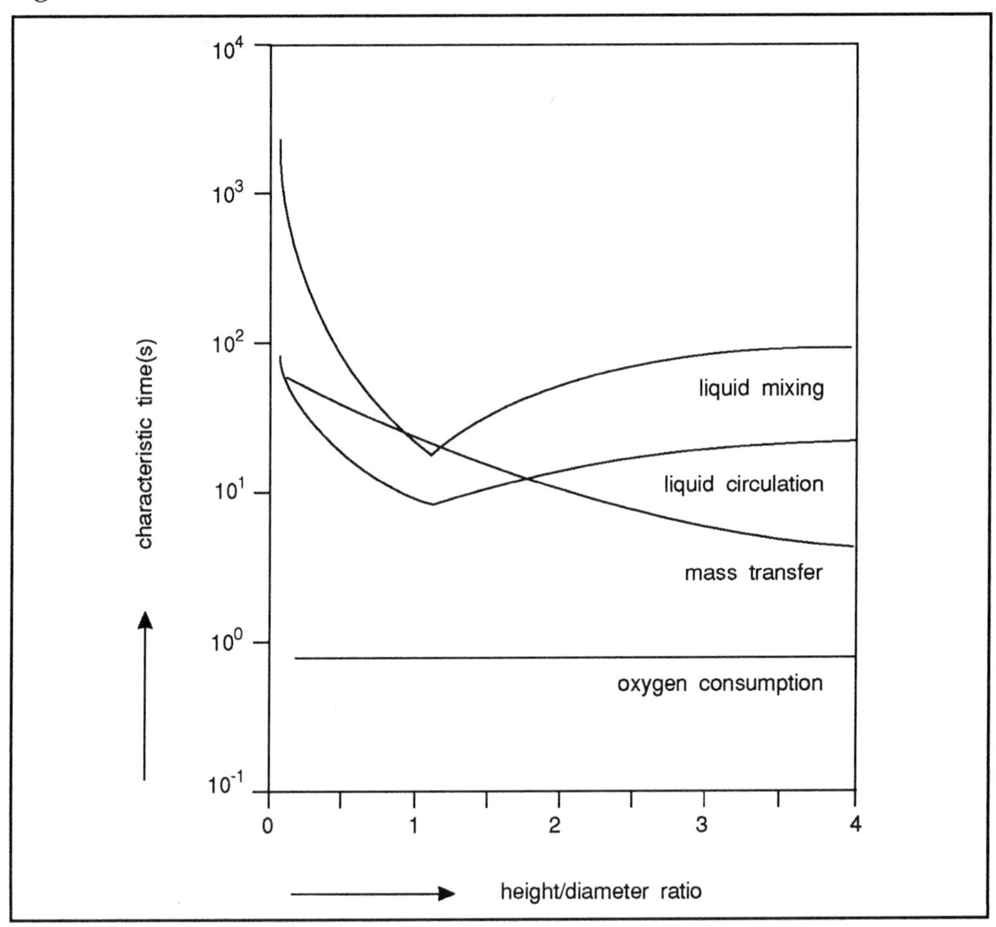

Figure 3.8 Estimated values of the characteristic times of fed-batch bakers yeast production as a function of the height-to-diameter ratio of bubble column reactor (V = 120 m^3; ϕ_g = 1 vvm). Data from Sweere 1988, PhD Thesis, TU Delft, The Netherlands.

Π From Figure 3.8, can you identify a range of H/D ratios in which the reactor performance is governed by a mixed regime? (Remember that the governing regime is the one with the longest characteristic time).

If H/D is < 0.3 or > 3 the reactor performance is governed by the liquid mixing: a pure regime. If H/D is > 0.3 and < 3 the reactor performance is ruled by both liquid mixing and mass transfer: a mixed regime. Note that H/D = 1 a discontinuity occurs due to the fact that, in the correlations used, the smallest values of the height and diameter are used. This is based on the fact that the macro-mixing of the liquid phase in the reactor is determined by the smallest dimension of the column. The characteristic times for mass transfer and oxygen consumption are independent of the height of the column. From the ratio of these times the conclusion can be drawn that oxygen limitation will always occur.

A similar analysis can be made for substrate concentration in the fermenter (Figure 3.9). This figure shows some characteristic times as a function of batch time for a fed-batch system. In these calculations the gas flow rate was 5 cm s^{-1} and the volume increased from 80 m^3 to 120 m^3 in 10 hours. Liquid circulation time and liquid mixing time increases, due to the volume increase. The time for substrate addition and substrate consumption decrease, due to an increase in the biomass concentration. From the ratio of liquid mixing time and liquid circulation time to substrate addition and consumption time, it can be concluded that substrate gradients will occur and that the intensity of these substrate gradients increases during the fermentation. A characteristic of the fed-batch process is the limitation of the added component. From the ratio of substrate consumption and addition time it can be concluded that substrate limitation probably will occur.

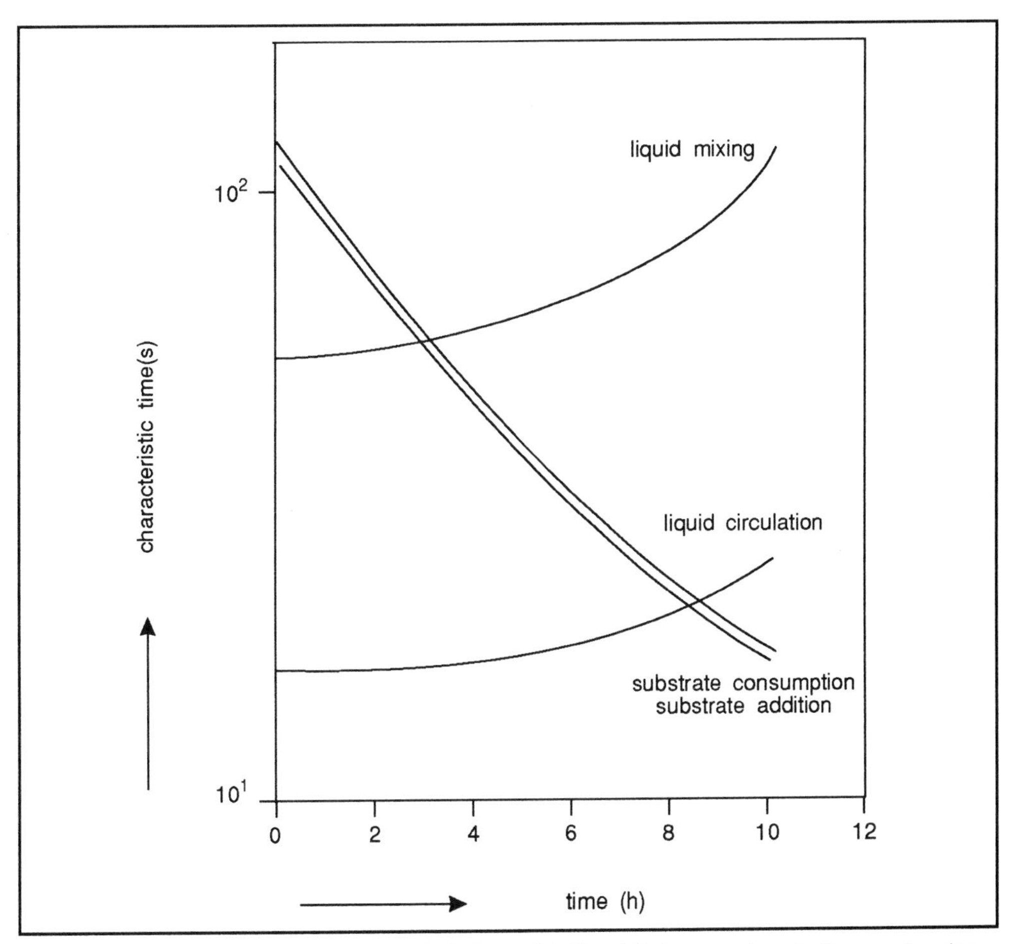

Figure 3.9 Estimated values of the characteristic times of fed-batch bakers yeast production as a function of batch time in a bubble column reactor (V = 80 - 120 m^3; at V = 120 m^3, H/D = 4, v_s = 0.05 m s^{-1}). Data from Sweere 1988, PhD Thesis, TU Delft, The Netherlands.

3.6 Conclusions

In this, and the previous chapter, we have described the different methodologies that can be used in scale up problems. You have probably already made you own judgement as to which of the methods are most useful. We have tried to pass our judgement as to which are useful and which are not, in the form of a diagram (Figure 3.10). It would be interesting to compare your judgement with ours.

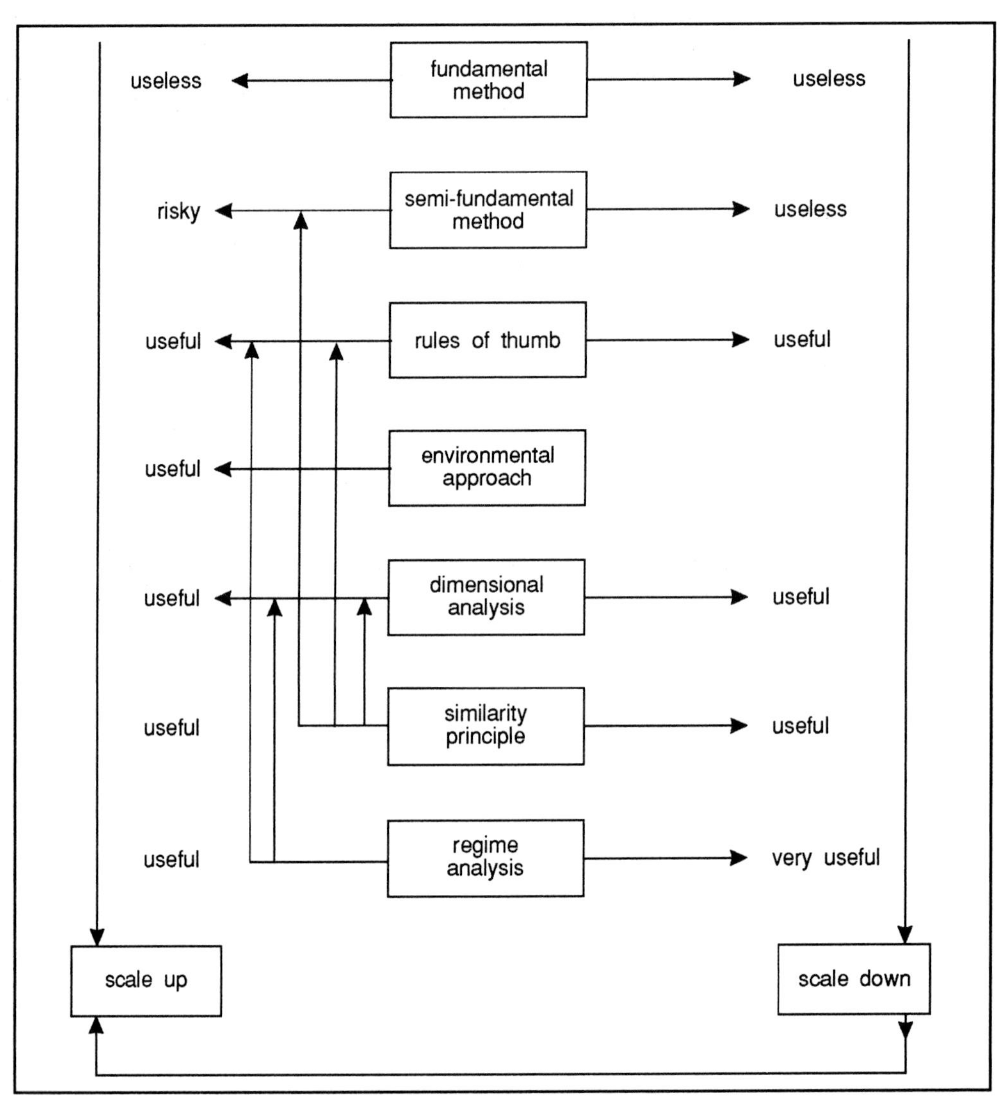

Figure 3.10 Comparative usefulness or otherwise of the methods used in scale-up (see text for details). The vertical arrows indicate that information from one type of analysis can be used to supplement other types of approaches.

In our opinion one of the most successful approaches is to use a scale down procedure based upon regime analysis. The scale down concept can be used both for the optimisation of existing processes and for the development of new bioreactors.

SAQ 3.4

1) The characteristic times for a reactor for liquid mixing (t_m), oxygen transfer (t_{OT}), oxygen consumption (t_{OC}), circulation time (t_C) are 10s, 11s 50s, 2.5s respectively a) is this reactor performance governed by a pure or mixed regime, b) is oxygen rate limiting?

2) A batch culture produces a viscous product. This viscous product changes a variety of characteristic times as shown in the table of data below. Comment on the rate limiting regime(s) at the various time intervals shown and whether or not gradients will be established in the reactor.

Characteristic times (s)

Time of batch culture (hr)	$t_{m,l}$	t_{OC}	$t_{OT,l}$	t_{hp}	t_{ht}
0	10	380	1	3000	40
20	10	340	3.0	1050	48
40	18	260	6.5	750	56
60	26	114	11	400	80
80	65	40 (at 85 h)	22	360	100
100	100	18 (at 96 h)	68	300	400

where: $t_{m,l}$ = liquid mixing time; t_{oc} = oxygen consumption time; $t_{OT,l}$ = oxygen transfer time into the liquid; $t_{h,p}$ = heat production time; $t_{h,t}$ = heat transfer time.

Summary and objectives

In this chapter we have extended your knowledge of the methods that can be employed in the design of bioreactors by examining the scale down strategy. This, we showed to be particularly valuable if coupled with a regime analysis approach. We illustrated this approach using a number of published examples.

Now that you have completed this chapter you should be able to:

- identify rate limiting mechanisms from supplied data relating to characteristic times;
- suggest experimental strategies which may enable relief of the rate limiting mechanism leading to process optimisation;
- identify the likelihood of gradients being created in reactors from supplied data relating to characteristic times;
- describe the usefulness or otherwise of the various methods used in scale up;
- describe examples in which the scale down approach has been used to design bioreactor processes.

Continuous flow stirred tank reactors

Continuous flow stirred tank reactors

4.1 Introduction

The term reactor can be applied to any physical, well defined space in which chemical or biological conversion reactions occur. A large variety of such spaces can be found that satisfy this definition. Natural reactors are those which either have developed over a long period of time (rivers, lakes) or occur spontaneously after a particular event (puddles after a rain shower). Man-made reactors can be either designed to function as a reactor (beer fermentation vessel, activated sludge plant) or can be designed to perform entirely different functions (irrigation canal, waste water collection mains, tubes to supply nutrient medium into a laboratory reactor). We can therefore identify a large variety of reactors, of different shapes, sizes and forms. In this, and subsequent chapters, we will examine the main types of man-made reactors.

The discussion presented here is written on the assumption that the reader has an understanding of the fundamental principles of transfer processes and process modelling and some knowledge of the basic modes of operation of bioreactors covered in the BIOTOL texts 'Bioprocess Technology: Modelling and Transport Phenomena' and 'Operational Modes of Bioreactors'. Despite this assumption, some reminders of these principles are incorporated into the text. The aim of this chapter is to apply these fundamental bioprocess engineering principles to continuous flow stirred tank reactors.

This will be done on the basis of applying balance equations and kinetic expressions to establish important relationships between biomass and product formation, substrate consumption and yield functions. This is a long chapter, but is followed by two quite brief chapters. The first (Chapter 5) examines batch reactors whilst in the second (Chapter 6) we will discuss plug flow reactors. These reactors all have one particular feature in common, the bioconversion process occur in suspension and in the continuous processes, biomass is lost from the reactor. In Chapter 7 we describe the effects of retaining all or part of the biomass in a reactor by using a cell retention system in which cells are retained by recirculation. In Chapter 8, we will discuss reactors in which cells are retained by sorption processes. These will be introduced on the basis of a prototype biofilm reactor and thereafter be extended into a discussion of the various fixed film reactors.

In this chapter we will first describe, in general terms, continuous flow stirred tank reactors (CFSTRs) before examining balance equations for these reactors. This will enable us to establish steady state equations for these reactors as well as kinetic relationships within these reactors. Based on these relationships we will be able to establish substrate consumption rates and to develop strategies to maximise substrate removal, biomass production and product formation rates. In most instances maximising substrate removal, biomass production and product formation are the primary objectives of the bioprocess engineer.

Critical to the success of a bioprocess is the need to maintain the biological system in a viable state. Thus a section is included on cell maintenance and decay.

4.2 General features of a continuous flow stirred tank reactor

The continuous flow stirred tank reactors (CFSTR also referred to as the continuous stirred tank reactor [CSTR] or the chemostat) is a reactor with a continuous inflow and a constant liquid volume (Figure 4.1).

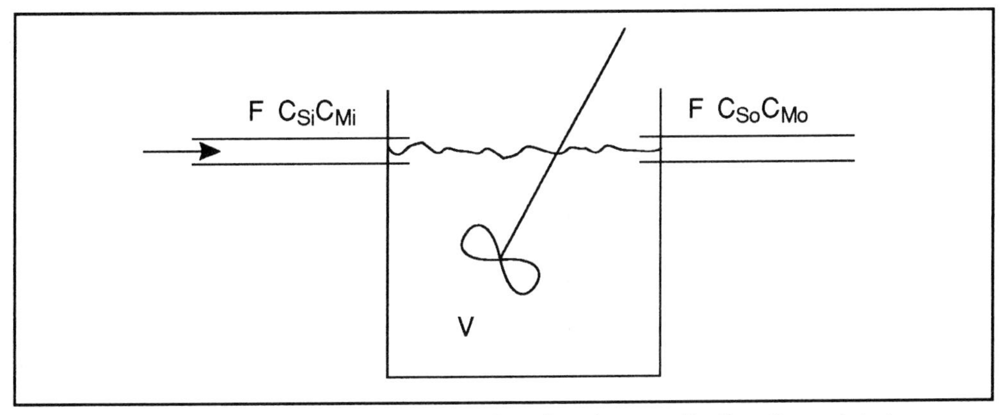

Figure 4.1 A CFSTR system showing inflow and outflow of a substrate s. F = flow, C_{Si} = substrate concentration in the inflow, C_{So} = substrate concentration in the outflow, C_{Mi} = biomass concentration in the inflow, C_{Mo} = biomass concentration in the outflow, V = volume of liquid in the reactor. Note that C_{So} = C_{Sl} where C_{Sl} = substrate concentration in the liquid in the reactor (see text for further details).

A basic principle of the CFSTR, as indicated by the words 'stirred tank', is the homogeneously mixed nature of the reactor contents. For the purposes of our discussion here, we will assume that:

- the concentration of a compound in the reactor is the same in each point of the reactor volume;
- the concentration of a compound in the reactor effluent equals the concentration in the reactor and;
- upon entering the reactor, compounds in the reactor influent are instantly homogeneously distributed.

Note that within this text we will use A and B to denote in general, compounds in the reactor.

The concept of constant volume implies that the reactor inflow equals the reactor outflow and that a change in the mass of compound A in the reactor is due to a change in the concentration of A in the reactor and not due to a change in the reactor volume.

An important characteristic of the CFSTR is its constant conditions. Once the system has reached the desired operational condition, it can be maintained that way as long as the environmental conditions, for example nutrient supply rate, remain the same.

If the objective of the process can be achieved with just one single operational condition, such as substrate removal (for example waste water treatment) or biomass production (for example baker's yeast production), continuous operation could well be the process of choice. However, if the objective of the process requires to interchange between two process conditions (for example in the production of penicillin in which a cell mass production step is followed by a penicillin production step), continuous operation becomes at least more complicated.

Examples of the CFSTR can be found primarily in environmental engineering. For example, the activated sludge tank, used in waste water treatment.

4.3 Balance equations for the CFSTR

4.3.1 Generalised relationships

The specific characteristics of CFSTRs can best be explained on the basis of mass balance equations. The general concept of balance equation is verbally represented by:

$$\boxed{\text{Rate of accumulation}} = \boxed{\text{Net rate of transport}} + \boxed{\text{Rate of transformation}} \qquad \text{(E - 4.1)}$$

Note that net rate of transport in = transport in - transport out

On the basis of Equation 4.1, a macrobalance for an arbitrary species A in a CFSTR can be derived from a microbalance as follows. Consider the reactor of Figure 4.2 with sides Δx, Δy and Δz. Assume Δx is much smaller than Δy. Further assume that the flow of mass into the reactor is only in the x-direction; flow enters the reactor only through side Δy Δz at $x = x$ while flow leaves the reactor only through side Δy Δz at $x = x + \Delta x$. This last assumption allows us to consider the problem one-dimensionally.

A mass balance in the volume element $\Delta x.\Delta y.\Delta z$ equals:

$$\Delta x \,.\, \Delta y \,.\, \Delta z \,.\, \frac{dC_A}{dt} = -\left[\underset{x = x}{n_{Ax}\Delta y \,.\, \Delta z} - \underset{x = x + \Delta x}{n_{Ax} \,.\, \Delta y \; \Delta z} \right] + R_A \,.\, \Delta x \,.\, \Delta y \,.\, \Delta z \qquad \text{(E - 4.2a)}$$

where:

C_A = mass concentration of species A in reactor ($M\ L^{-3}$)

n_{Ax} = flux of species A in the x-direction ($M\ L^{-3}t^{-1}$)

R_A = rate of reaction of species A per unit volume ($M\ L^{-3}t^{-1}$). R_A is positive in case of production, negative in case of consumption of compounds.

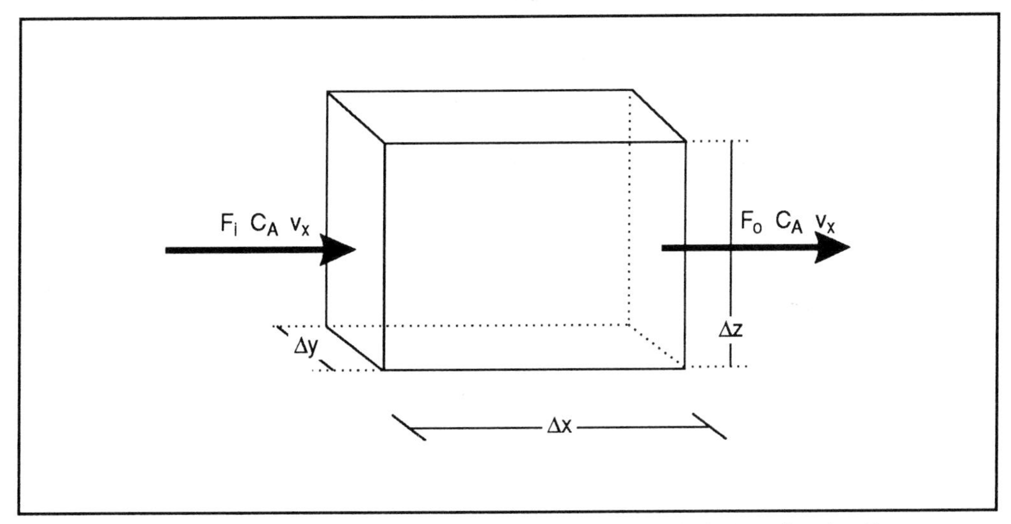

Figure 4.2 A differential element for a mass balance equation. v_x = velocity in the x direction, F = volumetric flow rate (see text for other symbols).

The flux n_{Ax} consists of a diffusive flux term, q_{Ax} and a bulk flux term, v_xC_A, where v_x = velocity in the x direction. In larger scale applications, the diffusive flux term is often negligible relative to the bulk flux term. Therefore, n_{Ax} can be considered to consist of the bulk flux term only. Substitution yields:

$$\Delta x \,.\, \Delta y \,.\, \Delta z \,.\, \frac{dC_A}{dt} + - \left[v_x \,.\, C_A \,.\, \Delta y \,.\, \Delta z \Big|_{x=x} - v_x \,.\, C_A \,.\, \Delta y \,.\, \Delta z \Big|_{x=x+\Delta x} \right] + \\ + R_A \,.\, \Delta x \,.\, \Delta y \,.\, \Delta z \qquad \text{(E - 4.2b)}$$

We can however modify this further.

A characteristic feature of the CFSTR is that the flow into the system equals the flow from the system. Therefore v_x (for x = x) equals v_x (for x = x + Δx). Moreover, $v_x\, \Delta y\, \Delta z = F_x = F$, where F is the volumetric flow rate. If you think about this for a moment you will see that it is valid, since $v_x\, \Delta y\, \Delta z$ is effectively velocity x surface area (= $\frac{\text{volume}}{\text{time}}$ = flow rate).

If we denote the concentration of A on entry into the element as C_{Ai} and on leaving as C_{Al} (since the concentration on leaving is because of the definition of the CSTR, equal to the concentration in the bulk liquid, and thus denoted as C_{Al}) then C_A (for x = x) is C_{Ai} and C_A (for x = x + Δx) is C_{Al}. Considering that Δx Δy Δz = volume V this modifies Equation 4.2b after substitution:

$$V \,.\, \frac{dC_A}{dt} = F \,.\, (C_{Ai} - C_{Al}) + R_A \,.\, V \qquad \text{(E - 4.3)}$$

Π Check the consistency of the units of Equation 4.3.

We will use L to denote a linear measurement, M to represent mass, t to represent time and the symbol [=] to denote 'has the units of'.

Thus $V \; [=] \; L^3 \; ; \; C_{Al} \; [=] \; M_A L^{-3} \; ; \; t \; [=] \; t^{-1}$

Thus $V \dfrac{d\,(C_{Al})}{dt} \; [=] \; L^3 M_A L^{-3} t^{-1} \; [=] \; M_A t^{-1}$

Likewise:

$F\,(C_{Ai} - C_{Al}) \; [=] \; L^3 t^{-1} M_A L^{-3} \; [=] \; M_A t^{-1}$, and

$R_A \; [=] \; M_A t^{-1}$

Therefore the units in Equation 4.3 are consistent.

SAQ 4.1

We have a CFSTR in which a compound is pumped in at a rate of 5 l h^{-1} and at a concentration of 1 g l^{-1}. The compound does not undergo any reaction in the vessel. What will be the steady state concentration of this compound in the reactor? (Use Equation 4.3 to prove your answer).

4.3.2 Relationships involving more than one species

Before we move on let us explain why we use the term C_M to denote biomass concentration whereas in some texts the term [X] is used to denote biomass concentration.

In the early days of microbiology, it was discovered that microbial substrate was consumed in a reactor through some conversion process and that as a result some material was produced. Because the precise identity of this material could not be determined with any accuracy, it was, in analogy with the unknown X in mathematics, referred to with X and called biological mass or biomass. Later, it was found that biomass consisted of microbial cells and sometimes, microbial product and thus X tends to have limited application. We have adopted more generalised forms to denote concentrations (such as C_A and C_B as these have wider application). As we are now aware, biological conversions involve more than one species: substrate S is converted to microbial cell mass M and/or to mass of microbial product P.

Balance equations for a system in which substrate S converted to microbial cell mass M and microbial product P (therefore three mass balance equations) are analogous to Equation 4.3:

For species S:

$$\underset{\text{accumulation}}{\text{rate of}} = \underset{\text{transport in}}{\text{net rate of}} + \underset{\text{biomass M}}{\text{rate used to produce}} + \underset{\text{product P}}{\text{rate used to produce}}$$

$$V\frac{d\,(C_{Sl})}{dt} = F\,(C_{Si} - C_{Sl}) + R_{SM}\,V + R_{Sp}\,V \qquad \text{(E - 4.4)}$$

Note our nomenclature: R_{SM} = rate of substrate consumption to produce biomass, R_{SP} = rate of substrate consumption to form product.

For species M:

(rate of accumulation) = (net rate of transport in) + (rate biomass produced)

$$V\,.\,\frac{d\,(C_{Ml})}{dt} = F\,.\,(C_{Mi} - C_{Ml}) + R_{MS}\,.\,V \qquad \text{(E - 4.5)}$$

For species P:

(rate of accumulation) = (net rate of transport in) + (rate of production of P)

$$V\,.\,\frac{d\,(C_{Pl})}{dt} = F\,.\,(C_{Pi} - C_{Pl}) + R_{PS}\,.\,V \qquad \text{(E - 4.6)}$$

where:

C_{Sl} = mass concentration of substrate in liquid phase ($M_S\,L^{-3}$)

C_{Si} = mass concentration of substrate in reactor influent ($M_S\,L^{-3}$)

R_{SM} = rate of substrate mass consumption for production of microbial cell mass ($M_S\,L^{-3}\,t^{-1}$)

R_{SP} = rate of substrate mass consumption for formation of microbial product mass ($M_S\,L^{-3}\,t^{-1}$)

C_{Ml} = mass of microbial cell mass in liquid phase ($M_M\,L^{-3}$)

C_{Mi} = mass of microbial cell mass in reactor influent ($M_M\,L^{-3}$)

R_{MS} = rate of microbial cell mass production from substrate ($M_M\,L^{-3}\,t^{-1}$)

C_{Pl} = mass of microbial product in liquid phase ($M_P\,L^{-3}$)

C_{Pi} = mass of microbial product in reactor influent ($M_P\,L^{-3}$)

R_{PS} = rate of product mass formation from substrate ($M_P\,L^{-3}\,t^{-1}$)

Note that the concentrations of biomass, substrate and product must be in the same form. Usually they are expressed in terms of the mass (moles) of carbon.

Let us examine these equations a little more thoroughly.

Equations 4.4, 4.5 and 4.6 are three differential equations describing the mass flows of three species in a continuous flow stirred tank reactor. Stated as they are these equations appear to describe independent processes, that is they do not seem to be linked in any way. However, in reality they are: substrate is consumed, microbial cell mass is produced, product mass is formed. The higher the rate of substrate consumption, the higher the rate of microbial cell mass production and/or product formation.

This linkage can be seen in the reaction rate terms R_{SM}, R_{MS} and R_{PS} in Equations 4.4 - 4.6. Starting with the microbial cell mass production term (Equation 4.5), R_{MS} was defined as the rate of microbial cell mass production from substrate. This reaction rate expression is generally considered to be somehow proportional to the concentration of the species M, according to:

$$R_{MS} = r_M \ . \ (C_{MI})^a \qquad \text{(E - 4.7)}$$

where:

r_M = proportionality constant called the specific rate of reaction ($M_M^{1-a} \ . \ L^{-3(2-a)} \ . \ t^{-1}$) and,

a = exponent (-)

In most experimental and practical conditions, the exponent of C_{MI} turns out to approach unity, in effect stating that the rate of microbial cell mass production is proportional to the cell mass concentration. This modifies Equation 4.7 to:

$$R_{MS} = r_M \ . \ C_{MI} \qquad \text{(E - 4.8)}$$

with r_M having units of t^{-1}

Π See if you can write down a simple reason why the term in Equation 4.7 should be = 1.

It may be easily conceivable that the rate of microbial cell mass production is proportional to the concentration of cell mass as in the process of growth (duplication), each micro-organism undergoes cell doubling; the more cells present the more cells will double.

Traditionally the specific rate of reaction or growth rate of cell mass (r_M in Equation 4.8) is indicated with μ:

$$R_{MS} = \mu \ . \ C_{MI} \qquad \text{(E - 4.9)}$$

where μ = specific growth rate of microbial cell mass (t^{-1})

SAQ 4.2

1) What will be the rate of microbial cell mass production when the specific growth rate is 1 h^{-1} and the biomass concentration is 0.1 mol biomass C l^{-1}? Make certain you use the same units. What have you assumed in your answer?

2) If the specific growth rate of the biomass described in 1) doubles what will happen to the value of R_{MS}?

For the product formation term (Equation 4.6) the situation is similar to the one for the microbial cell mass production term. R_{PS} is the rate of formation of product P. Total product formation is the result of microbial activity and is therefore proportional to the concentration of microbial cell mass M:

$$R_{PS} = r_P \cdot C_{Ml} \qquad \text{(E - 4.10)}$$

where r_P = specific rate of formation of microbial product ($M_P M_M^{-1} t^{-1}$).

The substrate consumption term consists of two rate expressions (R_{SM} and R_{SP} - see Equation 4.4). The first expression, R_{SM}, with units of ($M_S L^{-3} t^{-1}$), is negative and proportional to R_{MS} with units of ($M_M L^{-3} t^{-1}$) in the following way:

$$R_{SM} = -R_{MS} \cdot \frac{1}{Y_{MSo}} \qquad \text{(E - 4.11)}$$

where Y_{MSo} is the so-called microbial cell yield or growth yield and is defined as the observed ratio of the microbial cell mass produced and the substrate mass consumed:

$$Y_{MSo} = \frac{\text{microbial cell mass produced}}{\text{mass of substrate consumed}} \qquad \text{(E - 4.12)}$$

where Y_{MSo} = microbial cell yield ($M_M M_S^{-1}$).

Note that for reasons to be mentioned later, the yield as defined here is an observed yield and is indicated with the subscript 'o'.

Thus we can write Equation 4.11 in words as:

the rate of consumption of the substrate to produce biomass =

$$- \text{rate of formation of biomass} \times \frac{1}{\text{observed growth yield}}$$

SAQ 4.3

Assuming only cell mass is formed in a substrate conversion process, what would happen to the rate of consumption of substrate if the rate of microbial cell mass production is maintained at the same level, while the microbial cell yield halved?

Similarly the second expression R_{SP}, with units of ($M_S L^{-3} t^{-1}$) is expressing a consumption rate and, will also be negative and will be proportional to R_{PS} with units of ($M_P L^{-3} t^{-1}$) in the following way:

$$R_{SP} = -R_{PS} \cdot \frac{1}{Y_{PS}} \qquad \text{(E - 4.13)}$$

where Y_{PS} is the so-called microbial product yield and is defined as the ratio of the microbial product mass formed and the substrate mass consumed:

$$Y_{PS} = \frac{\text{mass of product formed}}{\text{mass of substrate consumed}} \qquad \text{(E - 4.14)}$$

where Y_{PS} = product yield ($M_P M_S^{-1}$)

Π Write beside Equation 4.13, a verbal version of this equation.

Combination of Equation 4.9 and Equation 4.11 and of Equation 4.10 and 4.13 gives:

$$R_{SM} = - \mu \cdot C_{Ml} \cdot \frac{1}{Y_{MSo}} \qquad \text{(E - 4.15)}$$

$$R_{SP} = - r_P \cdot C_{Ml} \cdot \frac{1}{Y_{PS}} \qquad \text{(E - 4.16)}$$

Thus the rate of substrate consumption to produce biomass is equal to the specific growth rate multiplied by the biomass concentration, and the reciprocal of the microbial cell yield (Y_{MSo}). Likewise, the rate of substrate consumption to form products is equal to the specific rate of product formation multiplied by the cell mass concentration, the volume of the reactor and the reciprocal of the microbial product yield (Y_{PS}). Note that the minus signs in Equations 4.15 and 4.16 indicate that substrate is being consumed.

Π Before reading on, take a piece of paper and substitute the rate expressions (Equation 4.9, 4.10, 4.15 and 4.16) into Equations 4.4, 4.5 and 4.6 to produce the mass balances equations for the three species CFSTR system. Then compare your results with those we have described in Equations 4.17 - 4.19.

For species S:

$$V \cdot \frac{d\,(C_{Sl})}{dt} = F \cdot (C_{Si} - C_{Sl}) - \mu \cdot C_{Ml} \cdot V \cdot \frac{1}{Y_{MSo}} - r_P\, C_{Ml}\, V \cdot \frac{1}{Y_{PS}} \qquad \text{(E - 4.17)}$$

For species M:

$$V \cdot \frac{d(C_{Ml})}{dt} = F \cdot (C_{Mi} - C_P) + \mu \cdot C_P \cdot V \qquad \text{(E - 4.18)}$$

For species P:

$$V \cdot \frac{d(C_{Pl})}{dt} = F \cdot (C_{Pi} - C_{Pl}) + r_P \cdot C_{Ml} \cdot V \qquad \text{(E - 4.19)}$$

From the yield coefficients in Equation 4.17 it can be seen that Equations 4.17 through to 4.19 are linked. Do SAQ 4.4 as this will prove the case to you.

SAQ 4.4

Verify the consistency of the units of Equation 4.17.

If you were successful in carrying out this SAQ, it will have shown you that each of the terms in Equation 4.17 is effectively a measure of the rate of change in substrate mass. We could put Equation 4.17 into the following words:

Rate of change in the mass of substrate in the reactor	=	Rate of change in the amount of substrate in the reactor as a result of flow into and out of the vessel	-	Rate of change in the amount of substrate in the reactor as a result of substrate consumption for growth	-	Rate of change in the amount of substrate in the reactor as a result of substrate consumption for product formation

Π Using this as an example, write down Equations 4.18 and 4.19 in verbal form.

The sort of expressions we hoped you will have written are:

For Equation 4.18:

Rates of change in the mass of biomass in the reactor	=	Rate of change in the amount of biomass in the reactor as a result of flow into and out of the vessel	+	Rate of change in the amount of biomass in the reactor as a result of growth

For Equation 4.19

Rate of change in the mass of product in the reactor	=	Rate of change in the amount of product in the reactor as a result of flow into and out of the vessel	+	Rate of change in the amount of product in the reactor as a result of product formation

4.4 The CFSTR concentration against time curves

Equations 4.17 through to 4.19 describe the rate of accumulation of species S, M and P in a CFSTR starting at time = 0. In general, time progressions of C_{S1}, C_{M1} and C_{P1} (graphs of carbon concentration C against time) are of sigmoidal (S) shape (Figure 4.3).

CFSTR processes generally start following a one-time addition of a small number of microbial cells (inoculum). Substrate is added continuously. At time = 0, $C_{S1} = C_{Si}$ while generally $C_{Mi} = 0$ and $C_{Pi} = 0$. Initially, the organisms need time to adapt to their new environment. In this 'lag phase' (Figure 4.3) virtually no growth occurs. After adaptation, the organisms start to grow. In the 'log phase' (between 3 and 8 days, Figure 4.3) the cell mass concentration and the product concentration increase. The substrate concentration in the reactor decreases proportionally. The log phase lasts until the substrate consumption rate in the reactor approaches the rate of substrate inflow rate into the reactor (F C_{Si}) as a result of which the substrate concentration in the reactor approaches 0. From that point onward (approx 9 day onwards - Figure 4.3), substrate inflow into the reactor can support only a certain growth rate of microbial cell mass and the process is in a steady state.

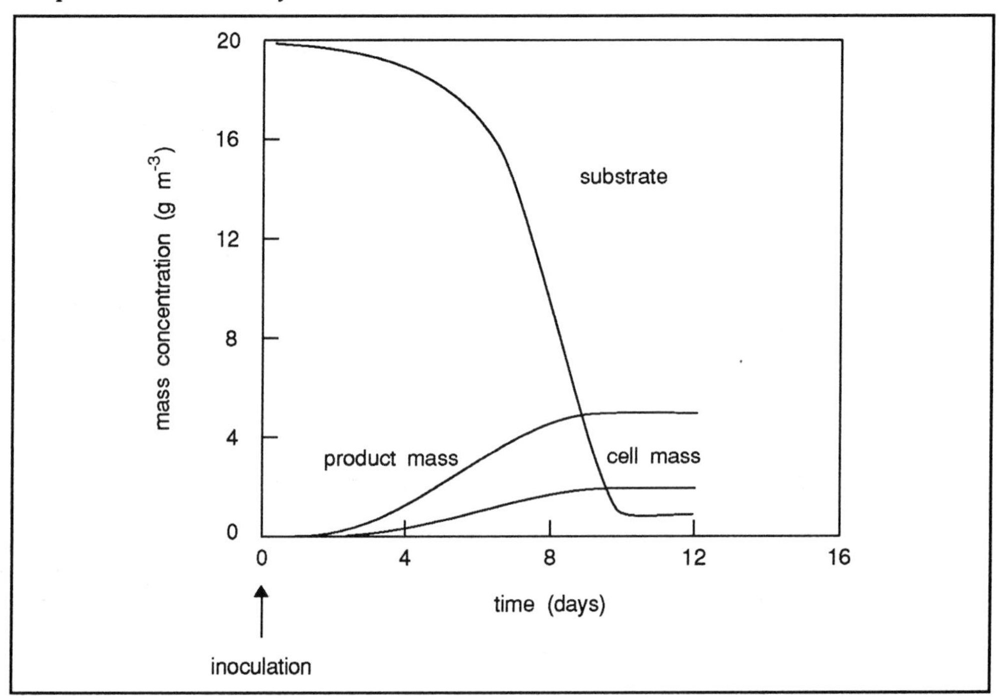

Figure 4.3 Example of the sigmoidal shape of curves representing time progression of substrate, cell and product mass in a CFSTR (from Equations 4.17, 4.18 and 4.19).

Π It might be quite helpful for you to mark on the 'lag phase', 'log phase' and 'substrate limitation phase' in Figure 4.3.

The lag phase described above was about 3 days long.

This is only a very approximate figure and the duration of the different phases vary enormously depending upon the system. In bioprocesses, we usually like to have a lag phase that is as short as possible.

Π List as many factors as you can which will tend to shorten the lag phase and then check your list with our comments below.

Many factors influence the duration of the lag phase. The following are however particularly important. Generally the smaller the inoculum the longer the lag phase. In principle, the larger the inoculum used, the more biomass we have to adapt to the new conditions; this shortens the lag phase. This is why when in industrial processes cultures are scaled up, we use a large inoculum. Also important is the age of the inoculum. If an old inoculum is used, the cells may be damaged (or even dead) and the culture takes a long while to recover and start growing. Thus generally the older the inoculum the longer the lag phase.

Also important is the extent of chemical and physical shock. If the inoculum is produced on the same medium and under the same physical conditions that it will find in the reactor the lag phase is usually short. The closer the match of these conditions, usually the shorter the lag phase. The greater the differences between conditions the greater the degree of adaptation the cells have to undergo and the longer the lag phase.

Π Re-examine Figure 4.3. How long does it take before the reactor illustrated becomes productive after inoculation?

If we take either substrate consumption or product formation as our measure of productivity then you will see from the data produced in Figure 4.3, that virtually nothing happens to either until day 3 - 4. Thus the operators are running an expensive process for no reward. This illustrates quite convincingly why we would like to have a minimum lag phase. It is only when we have a significant rate of bioconversion taking place that the process becomes productive.

4.5 CFSTR steady state equations

Equations 4.17 through to 4.19 describe both the steady state and the non steady state condition. Non steady state conditions last on average only a few hours to a few days while process engineering interest applies, generally, to the steady state part of the reaction.

For most engineering applications, for example the biotechnological production of penicillin or the treatment of waste water, only steady state conditions are of importance. Consequently, the accumulation terms in Equations 4.17 through to 4.19 can often be assumed zero resulting in steady state equations for the three species system:

For species S:

$$0 = F \,.\, (C_{Si} - C_{Sl}) - \mu \,.\, C_{Ml} \,.\, V \,.\, \frac{1}{Y_{MSo}} - r_P \,.\, C_{Ml} \,.\, V \,.\, \frac{1}{Y_{PS}} \qquad \text{(E - 4.20)}$$

For species M:

$$0 = F \,.\, (C_{Mi} - C_{Ml}) + \mu \,.\, C_{Ml} \,.\, V \qquad \text{(E - 4.21)}$$

For species P:

$$0 = F \,.\, (C_{Pi} - C_{Pl}) + r_P \,.\, C_{Ml} \,.\, V \qquad \text{(E - 4.22)}$$

Π In few processes will the accumulation term, dC/dt, ever get so close to zero that true steady state exists. Nevertheless, conditions of steady state are considered acceptable even if true steady state is not quite achieved. If in theory true steady state is indicated with 100%, at what percentage would you allow steady state conditions already applicable in practice? (You may not have sufficient experience to give a highly considered opinion, but have a go and then compare it with our comments).

In biological processes where steady is very much a condition of continuous variation around the average, steady state conditions are often assumed when the variations are less than about ± 30% around the steady state values. In other words, we are rather tolerant towards fluctuations in bioprocesses and any process that does not oscillate too violently around mean values and whose oscillations are within ± 30% of the mean, we assume to be in steady state.

Returning to Equation 4.20 - 4.22, the concentrations of microbial cell and product mass in the influent (C_{Mi} and C_{Pi}) are either zero or very small, relative to the concentrations of those components in the reactor effluent (C_{Ml} and C_{Pl}) and are, therefore, negligible. Dividing Equation 4.20 through to 4.22 by V and substituting D (= dilution rate $[t^{-1}]$ for F/V and rewriting with $C_{Mi} = C_{Pi} = 0$ gives:

For species S:

$$D \cdot (C_{Si} - C_{Sl}) = C_{Ml} \cdot \left(\frac{\mu}{Y_{MSo}} + \frac{r_P}{Y_{PS}} \right) \qquad \text{(E - 4.23)}$$

For species M:

$$D \cdot (C_{Ml}) = \mu \, C_{Ml} \qquad \text{(E - 4.24)}$$

For species P:

$$D \cdot (C_{Pl}) = r_P \cdot C_{Ml} \qquad \text{(E - 4.25)}$$

Try to do these substitutions yourself. From Equations 4.23 - 4.25, the following conclusions can be drawn: Dividing Equation 4.24 by C_{Ml} results in:

$$\boxed{D = \mu} \qquad \text{(E - 4.26)}$$

Equation 4.26 suggests that in a CFSTR the specific cellular growth rate, μ, equals the dilution rate. With D defined as F/V and V being constant, the specific cellular growth rate of a microbial species, μ, can be set by setting the flow rate F into the reactor. We have boxed in this relationship, it is an important one to remember.

Eliminating r_P and μ with Equations 4.26 and 4.25 from Equation 4.23 results in:

$$\boxed{C_{Si} - C_{Sl} = \frac{C_{Ml}}{Y_{MSo}} + \frac{C_{Pl}}{Y_{PS}}} \qquad \text{(E - 4.27)}$$

Equation 4.27 states that the difference between the substrate concentration in the reactor influent and effluent is used for the formation of the sum of the microbial cell mass and product mass. These two relationships (Equation 4.26 and 4.27) are fundamental and of great importance. Now attempt SAQ 4.5 which is related to these two equations.

SAQ 4.5

1) What will happen to the specific growth of a culture in a CFSTR when the dilution rate approaches zero?

2) What will happen when the dilution rate D exceeds the maximum rate of microbial growth, the so-called maximum specific growth rate (μ_{max})?

3) Formulate mathematically the term yield for a microbial organism not forming product, using the terminology of Equation 4.27.

4) Do the same for an organism not producing microbial cell mass.

4.6 Kinetic expressions

4.6.1 Monod relationship

Before being able to extract more quantitative information from the balance equations derived previously, the transformation rate terms (Equation 4.1) may be further specified. This means that the specific rate of growth of microbial cell mass, μ, and the specific rate of formation of microbial product, r_P, should be expressed in terms of known process parameters.

Various expressions for the specific growth rate μ have been considered extensively in the literature (see for example the BIOTOL text 'Bioprocess Technology: Modelling and Transport Phenomena'). One of the best known and most frequently used forms was introduced by Monod (1942 Annual Rev of Microbiol 3, 371) and is referred to as the Monod equation. Monod, in analogy to the Michaelis-Menten (Michaelis and Menten, 1913 Biochem. Zeitschrift 49,333) equation relating to the velocity of an enzyme catalysed process and substrate concentration, found that the rate of a microbial reaction (= microbial growth rate) is related to the concentration of the limiting reactant, according to:

$$\mu = \frac{\mu_{max} \cdot C_{S1}}{K_S + C_{S1}} \quad \text{(E - 4.28)}$$

where:

μ_{max} = the maximum specific growth rate of microbial cell mass (t^{-1})

K_S = the mass concentration of substrate S for which the growth rate (μ) equals half the maximum growth rate (μ_{max}) and is called the half saturation concentration ($M_S\ L^{-3}$).

Similarly, expressions have been developed to related the specific rate of formation of microbial product, r_P, to known process parameters. One relationship was given by Luedeking and Piret (1959 J of Biochem and Microb Tech and Eng - 1(4), 393), who indicated that the specific product formation rate could be partly related to the specific microbial growth rate (growth-associated) and partly was not related to growth (non-growth-associated):

$$r_P = k_{Pg} \cdot \mu + k_{Pn} \quad \text{(E - 4.29)}$$

where:

k_{Pg} = growth-associated product formation coefficient ($M_P M_M^{-1}$)

k_{Pn} = non-growth-associated product formation coefficient ($M_P M_M^{-1}\ t^{-1}$)

Π What is the value of k_{Pg}, if the product formation is independent of cellular growth rate?

You should have concluded that for organisms where product formation is independent of cellular growth rate, the term k_{Pg} is zero.

4.6.2 The CFSTR - steady state curve

Combining Equations 4.28 (the cell mass growth rate expression) and 4.29 (the product mass formation rate expression) with Equations 4.23 - 4.25 yields expressions for the relationship between substrate, cell and product mass concentrations at various dilution rates for a CFSTR at steady state. Let us work through some of these relationships together.

Rewriting Equation 4.28 and substituting D for μ gives an expression for the reactor substrate concentration. C_{SI}:

$$C_{SI} = \frac{D\,.\,K_S}{\mu_{max} - D} \qquad \text{(E - 4.30)}$$

Since $\mu = \dfrac{\mu_{max}\,C_{SI}}{K_S + C_{SI}}$ and $\mu = D$

Then $D = \dfrac{\mu_{max}\,C_{SI}}{K_S + C_{SI}}$ and $(K_S + C_{SI}) = \dfrac{\mu_{max}\,C_{SI}}{D}$

$$K_S = C_{SI}\left(\frac{\mu_{max}}{D} - 1\right)$$

$$K_S = C_{SI}\,\frac{\mu_{max}}{D}\left(1 - \frac{D}{\mu_{max}}\right)$$

$$C_{SI} = \frac{D\,K_S}{\mu_{max}\left(1 - \dfrac{D}{\mu_{max}}\right)}$$

$$C_{SI} = \frac{D\,.\,K_S}{\mu_{max} - D}$$

Similarly, rewriting Equation 4.23 after substitution of Equations 4.26 and 4.29 gives an expression for the reactor cell mass concentration:

$$C_{Ml} = \frac{D \cdot (C_{Si} - C_{Sl})}{\frac{D}{Y_{MSo}} + \frac{k_{Pg} \cdot D + k_{Pn}}{Y_{PS}}} \qquad \text{(E - 4.31)}$$

(You might like to try to do this derivation yourself).

Similarly, rewriting Equation 4.25 after substitution of Equations 4.26 and 4.29 gives an expression for the product mass concentration:

$$C_{Pl} = C_{Ml} \cdot \frac{k_{Pg} \cdot D + k_{Pn}}{D} \qquad \text{(E - 4.32)}$$

Equation 4.30 gives us the relationship between the steady state substrate concentration C_{Sl} and the dilution rate. Likewise, Equation 4.31 gives us the steady state biomass concentration and Equation 4.32 gives us the steady state product concentration.

Π Re-examine Equation 4.31. What would be the steady state biomass concentration in relation to the input and output substrate concentrations if no product other than biomass was produced?

The relationship is $C_{Ml} = (C_{Si} - C_{Sl}) \cdot Y_{MSo}$

Simply the product terms in Equation 4.31 disappear - so Equation 4.31 becomes:

$$C_{Ml} = \frac{D\,(C_{Si} - C_{Sl})}{\frac{D}{Y_{MSo}}} = (C_{Si} - C_{Sl}) \cdot Y_{MSo}$$

If you think about this carefully you will see that this is a common sense answer. The difference in the input substrate concentration and the output substrate concentration $(C_{Si} - C_{Sl})$ is the amount of substrate used per unit volume. If the growth yield (ie amount of cell biomass produced/amount of substrate used) is Y_{MSo}, then the amount of biomass produced = amount of substrate used x growth yield, ie:

$$C_{Ml} = (C_{Si} - C_{Sl})\, Y_{MSo}$$

Likewise Equation 4.32 can be simplified if the formation is not growth related (ie k_{Pg} = 0), then:

$$C_{Pl} = C_{Ml} \frac{k_{Pn}}{D}$$

Before we move on to what these types of relationships mean in practice, let us do some simple calculations.

SAQ 4.6

1) Assume no product other than biomass is produced from the substrate. If the input substrate concentration is 5 mol C m^{-3}, the output substrate concentration is 0.2 mol C m^{-3} and the growth yield is 0.5 mol biomass C/mol substrate C, what is the steady state biomass concentration?

2) a) If the steady biomass concentration is 10 mol biomass C m^{-3}, the rate of product formation is not growth related and has a value of 0.1 mol product C (mol biomass C)$^{-1}$ h^{-1}, what will be the steady state product concentration be when the dilution rate is 1 h^{-1}?

b) Using the same conditions as in a), except that the dilution rate is 0.5 h^{-1} what will be the steady state product concentration? Assume the steady state biomass concentration remains the same.

Let us now return to Equation 4.30-4.32 and examine what exactly happens in practice.

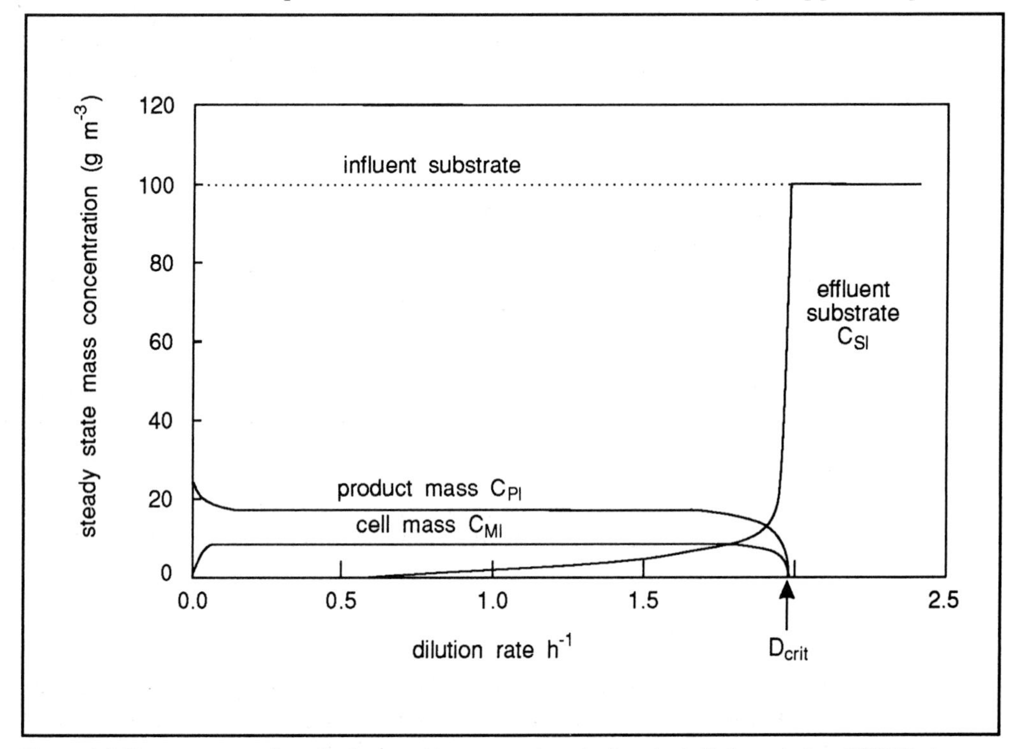

Figure 4.4 Mass concentration of substrate, biomass and product against dilution rate in a CFSTR at steady state (see text for details and parameter values). D_{crit} = critical dilution rate.

Figure 4.4 shows the relationship between dilution rate and substrate, cell and product mass concentrations in a CFSTR at steady state based on Equations 4.30 - 4.32 and on the following parameter values:

$\mu_{max} = 2\ h^{-1}$; $K_S = 1.4$ g C m^{-3}; $C_{Si} = 100$ g C m^{-3}; $Y_{MSo} = 0.19$ g C (g C)$^{-1}$; $Y_{PS} = 0.27$ g C (g C)$^{-1}$; $k_{Pg} = 2.3$ g C (g C)$^{-1}$; $k_{Pn} = 0.05$ g C (g C)$^{-1}$ t^{-1}.

Note that in this case we have used grams (g) of carbon to describe biomass, substrate and product whereas in the earlier in text activities we used mol of carbon. This is entirely permissable providing we are consistent within any single calculation.

In order to maintain microbial growth in a CFSTR, the dilution rate D has to be equal to or smaller than the maximum specific cellular growth rate, μ_{max} the value of D at which μ equals μ_{max} is referred to as the critical dilution rate D_{crit}. If D exceeds D_{crit}, microbial cell mass will not have sufficient time to reproduce before being discharged with the effluent and cell washout will occur. The microbial mass in the reactor will reduce and eventually disappear altogether.

Π Is the steady state biomass concentration in a CFSTR, greatly influenced by dilution rate? (Use Figure 4.4 to help you answer this).

Over a large range of dilution rates, there is very little effect on steady state biomass concentration. It is only at the extremes of dilution rate (D approaching zero or D_{crit}) that biomass is affected to any great extent. Note however that the exact relationships between C_{SI}, C_{PI} and dilution rate depend on the affinity the micro-organism has for the substrate and on the maintenance requirements of the organism. We will deal with these in later sections.

Likewise, effluent substrate concentrations are more or less constant and very low over an extensive part of the dilution rate range. Product concentrations are much more variable and depend upon the extent to which product formation is linked to growth. If it is substantially linked to growth, then the type of plot shown in Figure 4.4 is obtained. If it is independent of growth, then a different type of plot would result.

Π Draw a plot for such a growth independent product onto Figure 4.4. Use your experience of SAQ 4.6, 2a) and 2b) and Equations 4.30 and 4.32 to help you do this.

The sort of line we would anticipate you would draw would be to have a product mass line that starts high at low values of D and which declines as D increases since:

$$C_{PI} = C_{MI} \frac{k_{PA}}{D}$$

Again, however, the exact shape of the curve will depend upon the parameter values. Furthermore, we have assumed that the substrate would not become limiting. In practice substrate often becomes rate limiting.

4.6.3 Substrate consumption rate

We begin by reminding you of Equation 4.23.

$$D\ (C_{Si} - C_{SI}) = C_{MI} \cdot \left(\frac{\mu}{Y_{MSo}} + \frac{r_P}{Y_{Ps}} \right)$$

The left side of Equation 4.23 (the rate of substrate into minus the rate of substrate out of the reactor) is referred to as the rate of substrate removal or the substrate consumption rate:

$$R_S = D \cdot (C_{Si} - C_{Sl}) \qquad \text{(E - 4.33)}$$

where R_S = rate of substrate removal ($M_S\ L^{-3}\ t^{-1}$)

R_S provides an indication of the overall activity of the cell mass in the reactor. R_S relates to the end result of substrate removal without reference to how that end result was obtained; that is, how much biomass and product formation were involved in the consumption of the substrate.

Dividing by C_{Ml} defines the substrate removal rate per unit cell mass:

$$r_S = \frac{D \cdot (C_{Si} - C_{Sl})}{C_{Ml}} \qquad \text{(E - 4.34)}$$

where r_S = the specific rate of substrate removal ($M_S\ M_M^{-1}\ t^{-1}$).

Note that the symbol for the substrate removal rate R_S is **capitalised** while the symbol for the specific substrate removal rate, r_S is lower case (**not capitalised**).

SAQ 4.7

1) In analogy of the definition of (specific) substrate consumption rate, define the microbial cell mass production rate and the specific microbial cell mass production rate in a CFSTR.

2) Similarly, define the (specific) rate of formation of microbial product mass in a CFSTR.

Dividing Equation 4.23 by C_{Ml} and substituting r_S for $D\ (C_{Si} - C_{Sl})/C_{Ml}$ yields:

$$r_S = \frac{\mu}{Y_{MSo}} + \frac{r_P}{Y_{PS}} \qquad \text{(E - 4.35)}$$

We may interpret μ/Y_{MSo} as the specific rate of substrate consumption for cell mass production and r_P/Y_{PS} as the specific rate of substrate consumption for product formation. Equation 4.35 shows that r_S necessarily equals the sum of the two specific substrate consumption terms each corrected by the respective yield expression.

SAQ 4.8

If the specific rate of substrate removal is 12 h^{-1} at steady state in a CFSTR operating at a dilution rate of 0.5 h^{-1} and the specific rate of product formation is 1 h^{-1}, what is the product yield if the growth yield is 0.5 g biomass C (g substrate C)$^{-1}$?

4.7 Maximising substrate removal, cell mass production and product formation rate

In any large scale biological conversion, optimising the output rate is a prime concern. In a biotechnological (production) process that means pursuing the highest output of mass of cells or product per unit time. In an environmental (removal) process, that means trying to oxidise the largest mass of organic matter per unit time. In both cases, the discussion centres around maximising the output in terms of mass per time, ie maximising the output rate.

Using a CFSTR, this translates for a production process into a need for a high reactor cell or product mass concentration in combination with a high dilution rate. We hope to achieve the largest possible difference between the rate of substrate mass flowing into and out of the reactor.

Equation 4.33 defined the substrate consumption rate as the product of the dilution rate and the difference between the substrate concentrations in influent and effluent.

$$R_S = D \cdot (C_{Si} - C_{Sl}) \qquad \text{(Equation 4.33)}$$

Similarly, the cell mass production rate can be defined as the product of the dilution rate and the cell mass concentration (assuming $C_{Mi} = 0$) and the product mass formation rate as the product of the dilution rate and the product mass concentration (assuming $C_{Pi} = 0$). Thus:

$$R_M = D \cdot C_{Ml} \qquad \text{(E - 4.36a)}$$

$$R_P = D \cdot C_{Pl} \qquad \text{(E - 4.36b)}$$

C_{Sl} is expressed in terms of the dilution rate, D, in Equation 4.30 $\left(C_{Sl} = \frac{D \cdot K_s}{\mu_{max} - D} \right)$.

Substitution of Equation 4.30 in Equation 4.33 results in:

$$R_S = D \cdot \left[C_{Si} - \frac{D.K_S}{\mu_{max} - D} \right] \qquad \text{(E - 4.37)}$$

The point of maximum output rate, D_{max} is defined by: $dR_S/dD = 0$, $dR_M/dD = 0$, and $dR_P/dD = 0$ and is given by:

$$D_{max} = \mu_{max} \cdot \left[1 - \left[\frac{K_S}{K_S + C_{Si}} \right]^{\frac{1}{2}} \right] \qquad \text{(E - 4.38)}$$

Steady state values of R_S, R_M and R_P are plotted as a function of the dilution rate in Figure 4.5 using the following parameter values: $\mu_{max} = 2\ h^{-1}$, $K_S = 1.4$ g C m^{-3}, $C_{Si} = 100$ g C m^{-3}, $Y_{MSo} = 0.19$ g C (g C)$^{-1}$, $Y_{PS} = 0.27$ g C (g C)$^{-1}$, $k_{Pg} = 2.3$ g C (g C)$^{-1}$, $k_{Pn} = 0.05$ g C (g C)$^{-1}$ t^{-1}. Using these parameter values the dilution rate for maximum or consumption rates is at approximately $D = 1.7\ h^{-1}$.

Expressions similar to the one in Equation 4.38 are derived when substituting Equation 4.31 into Equation 4.36b or Equation 4.32 into Equation 4.36a. You might like to do these substitutions yourself to relate R_M and R_P to dilution rates.

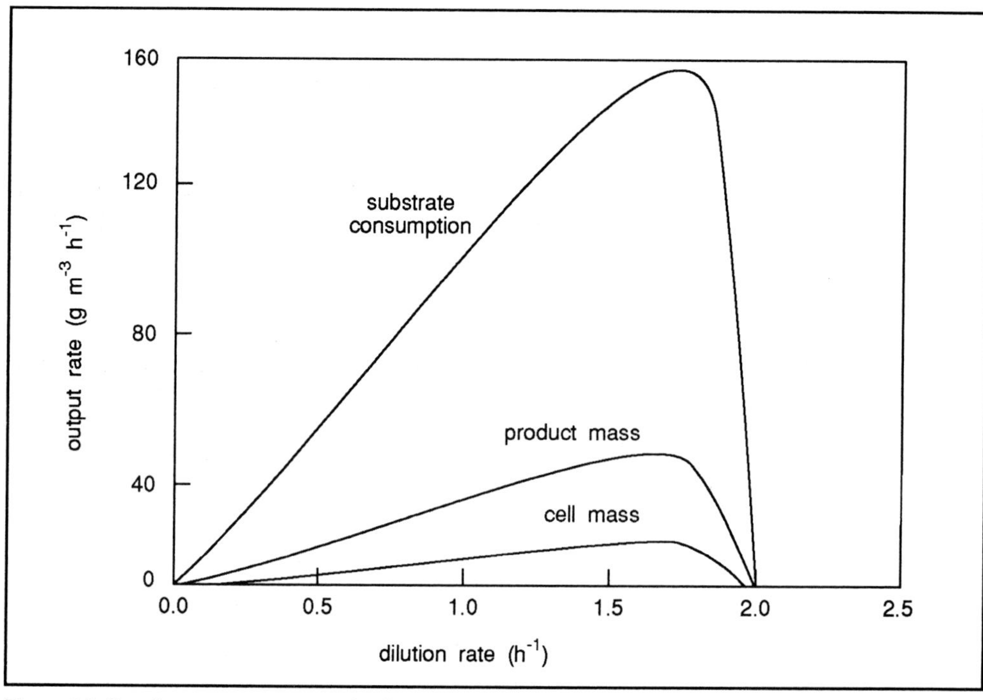

Figure 4.5 Steady state output rates of cell and product mass and consumption rate of substrate as a function of dilution rate.

Π The mathematic solution of the form $dR_S/dD = 0$ reads:

$$D_{max} = \mu_{max}\left[1 \pm \left[\frac{K_s}{K_s + C_s}\right]^{\frac{1}{2}}\right]$$

Yet, the dilution rate providing the maximum output rate is given according to Equation 4.38. Why? (Note the difference of ± and - between the two equations).

The microbial growth rate (and, hence, the dilution rate) for maximum production (or consumption) has to be below the maximum specific growth rate (and hence, the critical dilution rate). Thus we have written this in the form of:

$$D_{max} = \mu_{max}\left[1 - \left[\frac{K_s}{K_s + C_s}\right]^{\frac{1}{2}}\right]$$

SAQ 4.9

1) Given that the maximum specific growth rate of an organism is $\mu_{max} = 1\ h^{-1}$ and its $K_S = 1\ g\ C\ m^{-3}$. If the input substrate concentration = 143 g C m^{-3}, what is the maximum output rate for this system in a CFSTR?

2) If the concentration of product at this maximum output rate is 10 g C m^{-3}, what is the maximum output rate of the product.

3) If the concentration of biomass at this maximum output is 40 g C m^{-3}, what is the cell mass formation rate at the maximum output rate?

4.8 Cell maintenance and decay

4.8.1 Cell maintenance

The term yield was introduced (Section 4.3.2) as the ratio of microbial cell mass produced to substrate mass consumed (Y_{MSo}), and as the ratio of product mass produced to substrate mass consumed (Y_{PS}).

The cell mass yield is specified as an observed yield, meaning the yield as determined as it was observed. This suggests that the true ratio might have been different at a different point in time. Two phenomena indeed give rise to a cell yield which is different from the true yield. These phenomena are cell maintenance and decay.

The microbial cell consumes energy for various reasons: for cell division, for cell growth, for cell motility, for the formation of cellular products and for the maintenance of cell functions (Figure 4.6).

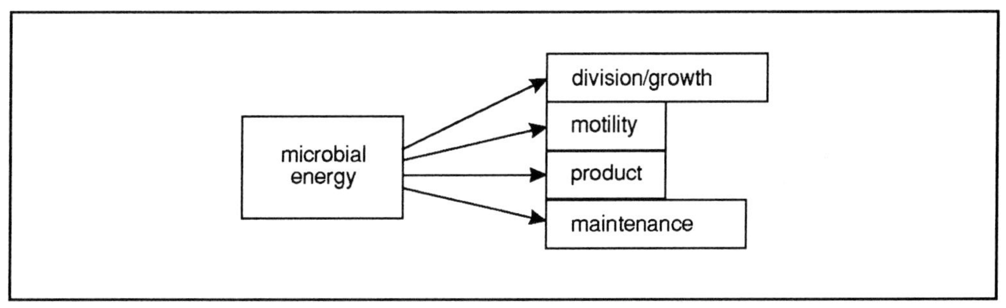

Figure 4.6 Energy usage by cells.

Maintenance energy is the energy necessary to maintain cell structure and integrity. It is not available for cell growth or biosynthesis. In an active microbial cell, the amount of energy consumed for cell maintenance purposes is negligible compared to the amount of energy used for growth, motility, etc. However, in conditions of low energy availability, the relative contribution of the maintenance energy increases, the energy usage for growth decreases.

Π Make a list of as many items as you can that may be considered as energy consuming and which perform a maintenance function.

There are many items you may have listed. Here we will describe the main ones. First, a cell has to maintain concentration gradients of chemicals. Most of the constituents in cells are at a much higher concentration than in the surrounding medium. For example Mg^{2+} concentrations are much higher in cells. The maintenance of such gradients costs energy. This energy expenditure, though essential for the production of biomass does not itself contribute to actual biomass. Likewise, the replacement of damaged cell constituents by new functional molecules costs energy but does not lead to any net production of biomass. Also the slow hydrolysis of cellular constituents such as ATP uncoupled to cellular processes leads to the production of heat but no functional use is made of this by cells. Thus this energy is effectively lost to the cell without any accumulation of biomass. Although these processes have no strict maintenance function, for energy and substrate consumption purposes we may treat them as such as they lead to no biomass accumulation.

The concept of maintenance energy requires a modification of the substrate mass balance equations through the addition of a maintenance term proportional to the microbial cell mass concentration, thus $r_m \cdot C_{Ml} \cdot V$ where r_m is the specific rate of maintenance energy consumption ($M_S M_M^{-1} t^{-1}$).

Π Consider Equation 4.17 and include maintenance as a substrate consumption term.

You should have written:

$$\frac{V\,d\,(C_{Sl})}{dt} = F \cdot (C_{Si} - C_{Sl}) - \mu \cdot \frac{C_{Ml}}{Y_{MSg}} \cdot V - r_P \cdot \frac{C_{Ml}}{Y_{PS}} \cdot V - r_m \cdot C_{Ml} \cdot V \qquad \text{(E - 4.39)}$$

Important - did you put in Y_{MSo} or Y_{MSg} in the second expression? You should have noted that in Equation 4.39, we did not use Y_{MSo} (observed growth yield) but instead Y_{MSg} (the true growth yield).

The yield coefficient, denoted by Y_{MSg} is the ratio of the concentration of cell mass produced and the concentration of substrate consumed for the production of this mass of microbial cells. In the definition Y_{MSo}, the denominator referred only to the difference in the substrate concentration in the inflow and outflow of the CFSTR therefore including possible other usages of substrate.

Assuming steady state conditions, dividing by V and C_{Ml} and substituting D for F/V results in:

$$\frac{D \cdot (C_{Si} - C_{Sl})}{C_{Ml}} = \frac{\mu}{Y_{MSg}} + \frac{r_P}{Y_{PS}} + r_m \qquad \text{(E - 4.40)}$$

or rearranged:

$$\frac{D \cdot (C_{Si} - C_{Sl})}{C_{Ml}} - \frac{r_P}{Y_{PS}} = \frac{\mu}{Y_{MSg}} + r_m \qquad \text{(E - 4.41)}$$

The left hand side of Equation 4.41 can be substituted by μ / Y_{MSo} through a combination of Equations 4.34 and 4.35.

Thus Equation 4.34: $r_S = D \dfrac{(C_{Si} - C_{Sl})}{C_{Ml}}$

and Equation 4.35: $r_S = \dfrac{\mu}{Y_{MSo}} + \dfrac{r_P}{Y_{PS}}$

Thus: $\dfrac{D\,(C_{Si} - C_{Sl})}{C_{Ml}} = \dfrac{\mu}{Y_{MSo}} + \dfrac{r_P}{Y_{PS}}$

Therefore: $\frac{D\,(C_{Si} - C_{Sl})}{C_{Ml}} - \frac{r_P}{Y_{PS}} = \frac{\mu}{Y_{MSo}}$ and $D\frac{(C_{Si} - C_{Sl})}{C_{Ml}} - \frac{r_p}{Y_{PS}} = \frac{\mu}{Y_{MSg}} + r_m$

(Equation 4.41)

Substituting Equations 4.26 ($D = \mu$) and dividing by D results in a relationship between the observed yield and the growth yield in a microbial conversion reaction:

$$\frac{1}{Y_{MSo}} = \frac{1}{Y_{MSg}} + \frac{r_m}{D} \qquad \text{(E - 4.42)}$$

Note that Y_{MSo} approaches Y_{MSg} when either r_m is small (r_m is an organism specific property dependent upon the environmental conditions) or when D is large.

Providing Y_{MSg} and r_m are constant at different dilution rates a plot of $\frac{1}{Y_{MSo}}$ against $\frac{1}{D}$ gives a straight line which cuts the Y_{MSo} axis at $\frac{1}{Y_{MSg}}$ while the slope is equal to r_m.

SAQ 4.10

The following steady state observed growth yields were obtained at the corresponding dilution rates using a CFSTR.

Dilution rates D (h^{-1})	Observed growth yields Y_{MSo} (g cell mass C (g substrate C)$^{-1}$)
1	0.495
0.5	0.490
0.25	0.480
0.12	0.460
0.06	0.428

Determine Y_{MSg} and r_m.

Thus we can determine the maintenance energy requirement. By conducting experiments with cells cultivated under different physical and chemical conditions we can, in principle, determine the conditions in which maintenance energy requirements are minimised.

4.8.2 Cell decay

Cell decay occurs when a microbial population is devoid of substrate. Some microbial cells will decay and literally break open releasing cell contents which may be used by other, still healthy cells as a source of nutrients. This obviously results in a net decrease in the size of the microbial population.

Cell decay can be accounted for in the mass balance equations through the addition of a decay term in the cell mass equation. The decay term is built up similarly as the biomass production term. Thus $r_d \cdot C_{Ml} \cdot V$ where r_d = the specific rate of microbial cell decay (t^{-1}).

Consider Equations 4.17 and 4.18 and add the decay term as a biomass consumption term.

We remind you of these equations.

For substrate (ie species S), we wrote:

$$V \cdot \frac{d(C_{Sl})}{dt} = F(C_{Si} - C_{Sl}) - \mu C_{Ml} \cdot V \cdot \frac{1}{Y_{MSo}} - r_P C_{Ml} V \frac{1}{Y_{PS}} \quad \text{(Equation 4.17)}$$

and for biomass M, we wrote:

$$V \frac{d(C_{Ml})}{dt} = F(C_{Mi} - C_{Ml}) + \mu C_{Ml} \cdot V \quad \text{(Equation 4.18)}$$

When including a biomass decay term in Equation 4.18, the yield coefficient in Equation 4.17 will describe the actual growth of cell mass, ie Y_{MSg} substitutes for Y_{MSo}.

Thus we can write:

For species S:

$$\frac{Vd(C_{Sl})}{dt} = F \cdot (C_{Si} - C_{Sl}) - \mu \cdot C_{Ml} \cdot V \cdot \frac{1}{Y_{MSg}} - r_P \cdot C_{Ml} \cdot V \cdot \frac{1}{Y_{PS}} \quad \text{(E - 4.43)}$$

For species M:

$$\frac{Vd(C_{Ml})}{dt} = F \cdot (C_{Mi} - C_{Ml}) + \mu \cdot C_{Ml} \cdot V - r_d \cdot C_{Ml} \cdot V \quad \text{(E - 4.44)}$$

Assuming steady state conditions, a negligible concentration of cell mass in the influent, substituting D for F/V and dividing by C_{Ml} alters Equation 4.43 to:

$$\frac{D \cdot (C_{Si} - C_{Sl})}{C_{Ml}} = \frac{\mu}{Y_{MSg}} + \frac{r_P}{Y_{PS}} \quad \text{(E - 4.45)}$$

and Equation 4.44 to:

$$\mu = D + r_d \quad \text{(E - 4.46)}$$

According to Equation 4.46, the specific growth rate, μ, in a CFSR is not necessarily equal to D but requires adjustment due to the influence of the specific rate of decay r_d. The influence of r_d is relatively minor at high dilution rates ($D >> r_d$) but becomes more apparent when D decreases.

We also remind you of the relationship between input and effluent substrate concentrations and biomass and product concentrations given in Equation 4.27.

$$C_{Si} - C_{Sl} = \frac{C_{Ml}}{Y_{MSo}} + \frac{C_{Pl}}{Y_{PS}}$$

Dividing Equation 4.27 by C_{Ml} gives the form:

$$\frac{C_{Si} - C_{Sl}}{C_{Ml}} = \frac{1}{Y_{MSo}} + \frac{C_{Pl}}{Y_{PS} \cdot C_{Ml}} \quad \text{(E - 4.47)}$$

or, after substitution of C_{Pl}/C_{Ml} by r_P/D from Equation 4.25 (since $D \cdot C_{Pl} = r_P \cdot C_{Ml}$):

$$\frac{C_{Si} - C_{Sl}}{C_{Ml}} = \frac{1}{Y_{MSo}} + \frac{r_P}{Y_{PS} \cdot D} \qquad \text{(E - 4.48)}$$

Substituting Equation 4.48 into Equation 4.45 results in:

$$\frac{D}{Y_{MSo}} = \frac{\mu}{Y_{MSg}} \qquad \text{(E - 4.49)}$$

Π Attempt to derive Equation 4.49 for yourself. If you cannot, check your approach with that given below (begin with Equation 4.45).

Since $$\frac{D\,(C_{Si} - C_{Sl})}{C_{Ml}} = \frac{\mu}{Y_{MSg}} + \frac{r_P}{Y_{PS}} \qquad \text{(E - 4.50)}$$

Then: $$\frac{\mu}{Y_{MSg}} + \frac{r_P}{Y_{PS}} = D\left(\frac{1}{Y_{MSo}} + \frac{r_P}{Y_{PS} \cdot D}\right) \text{(from Equation 4.48)}$$

$$\frac{\mu}{Y_{MSg}} + \frac{r_P}{Y_{PS}} = \frac{D}{Y_{MSo}} + \frac{D\,r_P}{Y_{PS}\,D} = \frac{D}{Y_{MSo}} + \frac{r_P}{Y_{PS}}$$

Thus $$\frac{\mu}{Y_{MSg}} = \frac{D}{Y_{MSo}}$$

Eliminating μ in Equation 4.49 using Equation 4.46 ($D = \mu + r_d$ and dividing by D) results in:

$$\frac{1}{Y_{MSo}} = \frac{1}{Y_{MSg}} + \frac{r_d}{Y_{MSg} \cdot D} \qquad \text{(E - 4.51)}$$

Apparently, the difference between Y_{MSg} and Y_{MSo} depends on the relative importance of the decay coefficient and/or the dilution rate. At high dilution rates or at low specific decay rates the observed yield approaches the growth yield.

SAQ 4.11

By determining the concentration of the cell mass and the substrate in a reactor, do we tend to over- or underestimate the microbial cell yield?

Comparison of Equation 4.42: $\left(\frac{1}{Y_{MSo}} = \frac{1}{Y_{MSg}} + \frac{r_m}{D}\right)$ with Equation 4.50

$$: \frac{1}{Y_{MSo}} = \frac{1}{Y_{MSg}} + \frac{r_d}{Y_{MSg} \cdot D}$$

indicates that, even though the reason for the difference between Y_{MSo} and Y_{MSg} is conceptually very different, the mathematical representations of both phenomena are comparable and that:

$$r_m = \frac{r_d}{Y_{MSg}} \tag{E - 4.52}$$

Maintenance results in less energy available for production of cell mass. Decay means a reduction of the cell mass after its formation. In both cases there is less cell mass produced per mass of substrate consumed than there would be if maintenance or decay had not be accounted for. As a result, the growth yield is higher than the observed yield, with the difference being a function of the dilution rate.

SAQ 4.12

1) If the specific rate of maintenance energy consumption is 1 g substrate C (g cell mass)$^{-1}$ h^{-1} and the true growth yield is 0.5g biomass C (g substrate C)$^{-1}$, what is the specific rate of microbial cell decay?

2) What will be the observed growth yield when the dilution rate is a) 1 h^{-1}, b) 0.5 h^{-1}.

When you have completed SAQ 4.12, read the response carefully because it has an important message.

4.9 Half saturation concentration

In the section on kinetic expressions, the rate expression derived by Monod was discussed. In the Monod model, the specific growth rate is expressed in terms of the reactor substrate concentration, C_{Sl}, the maximum specific growth rate, μ_{max} and the half saturation concentration, K_S. K_S is the substrate concentration at which the microbial growth rate is half the maximum microbial growth rate. K_S is related to the affinity of a microbial species for its substrate. The lower K_S, the larger the organism's ability to thrive at low substrate concentration for that specific substrate. An organism with a low value of K_S is also called oligotrophic, and an organism with a high K_S is called copiotrophic.

Although the half saturation concentration may appear to be another modeller's tool to translate microbial processes into mathematics, K_S also has a very practical significance. This is best explained in terms of microbial competition.

When, in a mixed culture, two species compete for the same substrate, one species may dominate the other in one part of the substrate concentration range, while that same species may be dominated at another substrate concentration (Figure 4.7). At high substrate concentrations (Figure 4.7: $C_{Sl} > C_{Sc}$ where C_{Sc} is the substrate concentration at which microbial co-existence is possible), the organism with the highest maximum specific growth rate, μ_{max} out competes the one with the lower μ_{max}. At low substrate concentrations (Figure 4.7: $C_{Sl} < C_{Sc}$) the value of K_S determines which organism is dominating: the organism with the lower K_S will out compete the one with the higher K_S. At the intersection of the two lines (Figure 4.7: $C_{Sl} = C_{Sc}$) the two species will theoretically co-exist.

The importance of K_S can also be shown in a different way. In Figure 4.8, steady state concentrations for C_{Sl} and C_{Ml} are shown for two organisms with the same maximum specific growth rate ($\mu_{max} = 2$ h^{-1}), but with different K_S values ($K_{S1} = 10$ g C m^{-3}, $K_{S2} = 1$ g C m^{-3}). The organism with a lower K_S value has a great ability to adsorb substrate even at low concentrations of substrate. This organism has a critical dilution rate D_{crit} at approx 2 h^{-1} while the organism with the higher K_S value washes out at approx 1.8 h^{-1}.

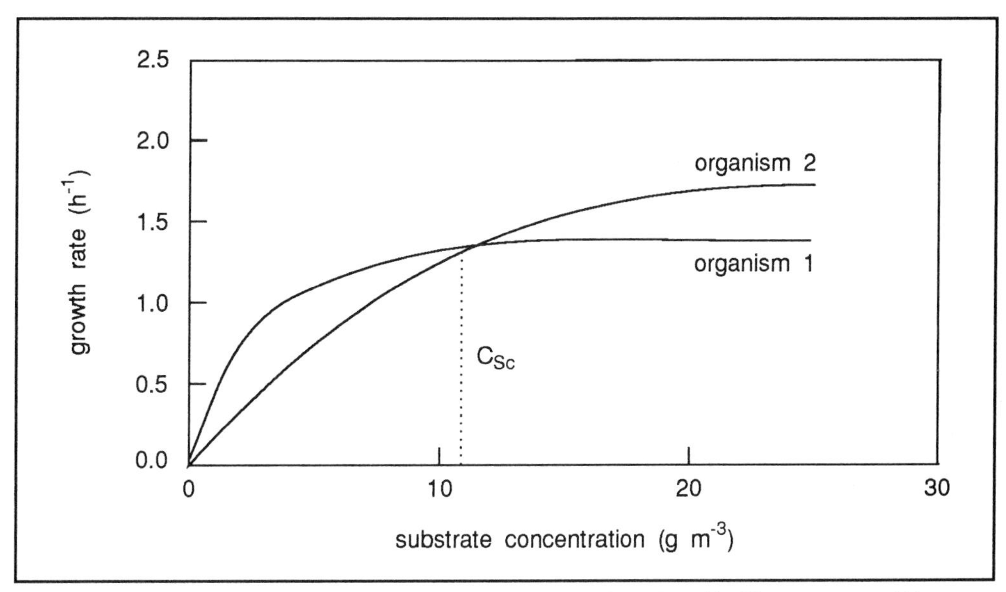

Figure 4.7 Comparison of Monod growth curves for two microbial species with different μ_{max} and K_S values (see text for discussion).

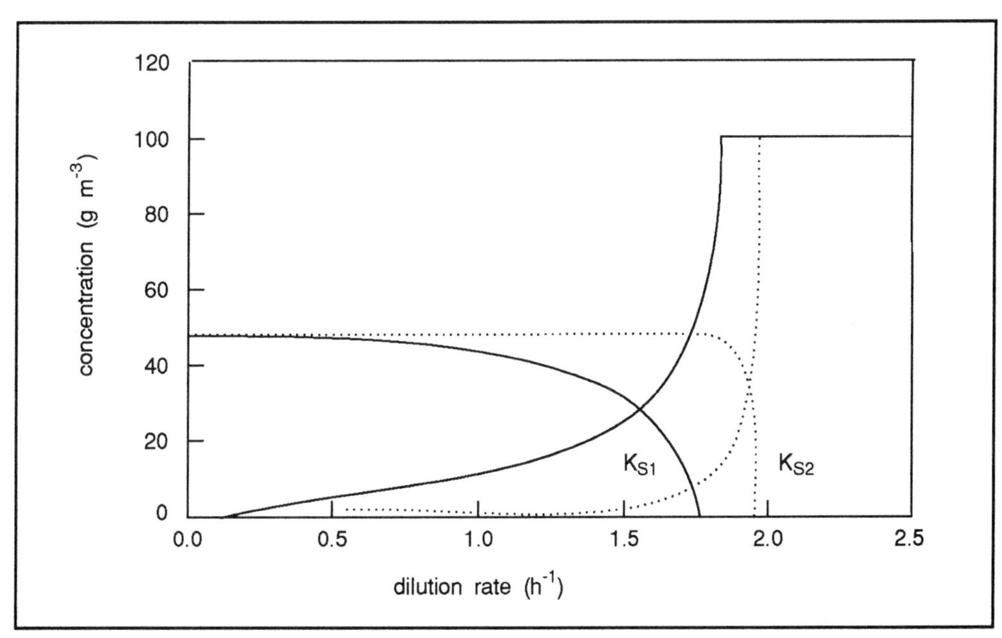

Figure 4.8 Steady state reactor concentrations as a function of dilution rate for 2 microbial species with different K_S values. See text for discussion.

SAQ 4.13

Below are listed the properties of two organisms I and II. From these properties, answer the questions set below.

Property	Organism I	Organism II
μ_{max}	1.2 h^{-1}	2.2 h^{-1}
K_S	0.01 g C m^{-3}	10.0 g C m^{-3}

1) Which organism is a copiotroph?

2) Which organism appears to be an oligotroph?

3) What will be the specific growth rates of the two organisms be at a substrate concentration of 10 g Cm^{-3}?

4) Which organism will dominate in a culture in which the substrate concentration is maintained at 0.05 g Cm^{-3}?

5) Which organism will dominate at a substrate concentration of 20 g C m^{-3}?

6) Is there a substrate concentration at which the two organisms can co-exist? If so, what is this concentration?

Summary and objectives

In this chapter we have examined continuous flow stirred tank reactors (CFSTR) with particular emphasis being placed on establishing relationships for biomass and product yield and substrate consumption. These are key features in both bioreactor design and performance measurements. Most bioreactors are run to produce biomass or product or both or to remove substrates. We began by discussing the principle features of CFSTRs before writing balance equations for this type of reactor. These in combination with kinetic expressions enable us to establish many important relationships between substrate consumption, biomass and product yield. In the final part of the chapter we explored the consequences of maintenance and cell decay on these relationships. We also established relationships for maximising substrate utilisation, biomass production and product formation.

Now that you have completed this chapter you should be able to:

- describe in general terms the operation of CFSTRs;
- calculate a wide variety of parameters such as steady state biomass concentrations, product concentration, substrate utilisation, observed growth yield, real growth yield and product yield from supplied data;
- use kinetic expressions, especially the Monod equation to calculate specific growth rates and substrate consumption rates from supplied data;
- use supplied data to determine specific maintenance energy consumption and specific cell decay rates;
- demonstrate the effects of maintenance energy consumption and cell decay on growth yields;
- determine maximum output rates from supplied data;
- use terms and units correctly.

Batch reactors

Batch reactors

5.1 Introduction

The batch reactor (BR) is the most frequently used type of reactor in biotechnological productions. Virtually all food processing, pharmaceutical and agricultural bioprocesses are carried out in batch reactors. Batch reactor applications in environmental processes are however relatively rare and very much limited to small scale applications.

One of the reasons for the difference in usage frequency between biotechnological and environmental technological applications of the BR is, no doubt, related to the fact that the BR is very well suited to operate under sterile conditions, something of great importance in biotechnology and often irrelevant in environmental processes. A BR is usually a closed, well mixed vessel (Figure 5.1). In the laboratory, the Erlenmeyer flask of a few millilitre is often used as a BR. In industrial applications BR's can have a volume of over 100 m^3.

We will begin this chapter by giving a brief description of batch reactors and by describing the batch process cycle including a discussion of the batch growth curve. We will then write balance equations for these types of reactors before moving onto a description of fed batch reactors.

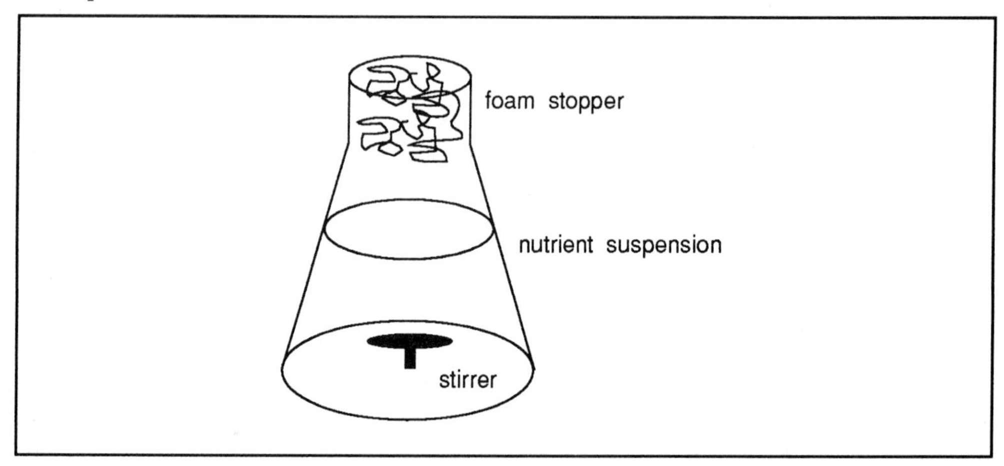

Figure 5.1 Batch reactor as used in the laboratory.

5.1.1 Batch reactors are usually regarded as closed systems

Operational intricacies are best related with the description of a full batch process cycle. Before inoculation, the BR contains a certain volume of nutrient in suspension. After inoculation, the process is left untouched, ie no material is added to or removed from the reactor. However, the reactor might be aerated, is virtually always stirred and process control, eg pH control, might be applied. Batch reactors are often referred to as closed systems.

The term 'closed' refers to the fact that material can neither enter nor leave the reactor. In that sense, the use of this term is often not entirely accurate because in aerobic processes, oxygen is allowed into the reactor (in laboratory conditions, through an air filter of cotton or foam) and carbon dioxide allow to leave. Obviously if a pH control system is used this will also lead to some additions to the reactor. Thus, in many cases batch reactors should not be regarded strictly as closed systems.

5.2 Batch growth curve

The biological conversion process in a BR usually proceeds through a series of phases (Figure 5.2).

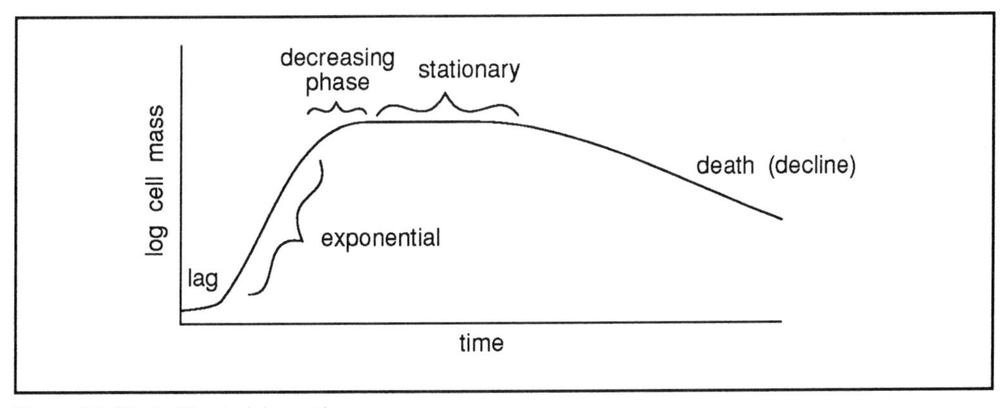

Figure 5.2 Typical bacterial growth curve.

After inoculation, micro-organisms need time to adjust to the new environment, ie to prepare the enzymatic machinery for the specific cellular functions involved in the process of growth. Virtually no growth occurs in this so called lag phase although it is a period of intense metabolic activity. Once adjusted to their environment, the organisms start to grow and, with all necessary growth factors plentifully available, growth occurs in an exponential (logarithmic) fashion. Hence, this phase is called the exponential or logarithmic growth phase. Growth in the log phase is referred to as balanced growth, defined as growth in which every extensive property of the growing system increases by the same factor.

Π Define the terms extensive and intensive properties. Identify three features of a microbial culture which might be regarded as extensive properties.

You may not have come across the terms extensive and intensive properties before so do not worry if you could not answer this. Intensive properties are all related to a system quality. These properties are not additive. Extensive properties relate to the system quantities and therefore are additive. Table 5.1 gives a summary of the features of these types of properties. Thus the extensive properties you could have used were such features as mass of cells present, mass of DNA or protein and so on. Colour and turbidity are examples of intensive properties.

Intensive Properties	Extensive Properties
- relates to system quality	- relates to system quantities
- cannot be added up	- can be added up
- no balance possible	- balance allowed
Examples	Examples
- temperature, pressure	- volume, mass
- colour, concentration	- energy, electric charge

Table 5.1 Features of intensive and extensive properties.

Returning to our description of the batch growth curve, during logarithmic (balanced) growth, each extensive property increases by the same factor and the composition of cells remains constant.

At some point, however, one or more nutrients start to become depleted and, hence, growth-limiting. As a result, growth first slows down (decreasing phase) and then virtually ceases, the beginning of the stationary phase. After growth has halted for some time the process of cell death and cell lysis sets in. Cell lysis is the breaking open of the cell membrane and the subsequent release of cell contents into the medium. Cell contents may provide nutrients for intact cells. Therefore, cellular growth rate in the stationary phase is temporarily balanced by the rate of cell lysis. Cell lysis reduces cell numbers until no living cells are left. This final stage is called the death phase (or decline phase).

It should be clear from the preceding description of the growth phase, that the batch reactor environment can be characterised as a continuously changing environment. It is an environment in which nutrient concentrations, physiological state of the microbial cell and mass of intact cells differ from one moment to another.

This fact provides the batch process with an important edge over constant environment processes (eg CFSTR). For example, cell mass may be produced most effectively when the cells are in the log growth phase, while product formation rates may be highest when the growth of cells has virtually stalled. For processes in which the trophophase does not coincide with the idiophase (see below) the batch reactor is well suited and often the only real possibility.

We have introduced some terminology here so let us explain them.

A microbial product formed in association with cell growth is called a primary metabolite, a product formed independent of cell growth is called a secondary metabolite.

Products are referred to by a variety of names, ie Extracellular Polymeric Substances (EPS), glycocalix and extracellular products while the word exopolymer is also used.

Idiophase refers to the phase of product formation. Trophophase means cell production phase. The idiophase and trophophase may or may not coincide.

SAQ 5.1

Below are the growth cycles of batch cultures of two organisms. Examine these carefully and answer the following questions. Products A and B are extracellular products (Note that these figures are somewhat stylised).

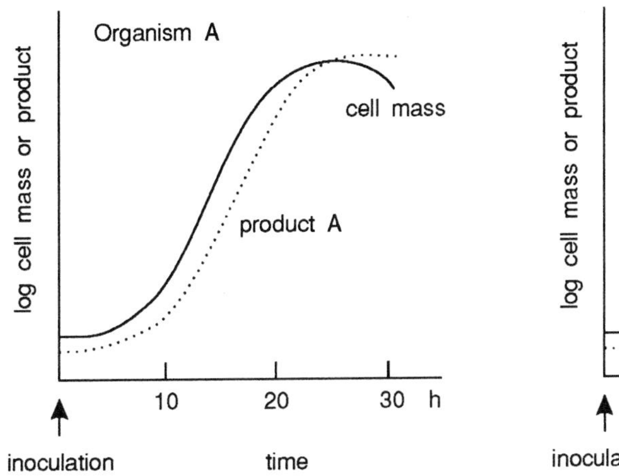

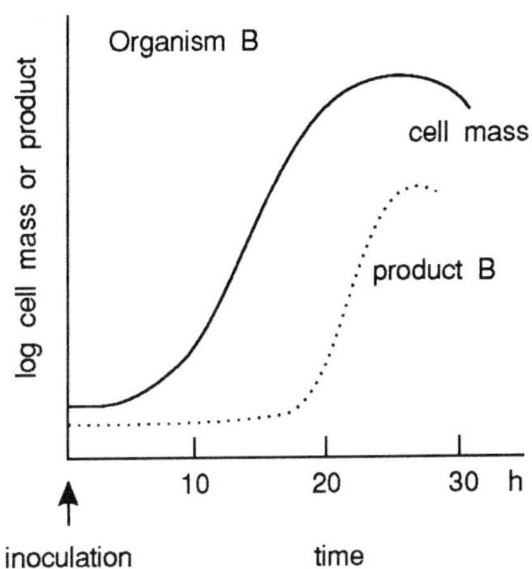

1) What is the duration of the idiophase in each culture?

2) What is the duration of the trophophase in each culture?

3) Which product is probably a primary metabolite and which a secondary metabolite?

4) For the production of which compound would a batch reactor be better than a CFSTR?

SAQ 5.2

Complete the following table.

	CFSTR	Batch reactor
Is it open or closed?		
Is the environment constant or varying?		
Is it operationally complex or simple?		

5.3 Balance equations for the BR

For a three species system of S $\rightarrow$ M + P. Mass balance equations for the batch reactor (BR) can be derived from those for the CFSTR by omitting the transport terms. Because part of the batch processes may occur at low substrate concentrations, a maintenance term should be included.

We remind you of some of the relevant equations for CFSTRs.

In a CFSTR, for species S:

$$V \frac{d(C_{Sl})}{dt} = F(C_{Si} - C_{Sl}) - \mu C_{Ml} . V . \frac{1}{Y_{MSo}} - r_P . C_{Ml} . V . \frac{1}{Y_{PS}} \quad \text{(Equation 4.17)}$$

For species M:

$$V . \frac{d(C_{Ml})}{dt} = F(C_{Mi} - C_{Ml}) + \mu . C_{Ml} . V \quad \text{(Equation 4.18)}$$

For species P:

$$V \frac{d(C_{Pl})}{dt} = F(C_{Pi} - C_{Pl}) + r_P . C_{Ml} . V \quad \text{(Equation 4.19)}$$

When we considered a situation in which part of the substrate in the CFSTR was used for providing maintenance energy, we wrote:

$$V \frac{d(C_{Sl})}{dt} = F(C_{Si} - C_{Sl}) - \mu \frac{C_{Ml}}{Y_{MSg}} V - r_P \frac{C_{Mi}}{Y_{PS}} V - r_m . C_{Ml} . V \quad \text{(Equation 4.39)}$$

These equations may be applied to BRs. From Equations 4.17 - 4.19, the BR mass balance equations for a three species system can be derived for $F = 0$ and including the maintenance term of Equation 4.39:

For species S:

$$V . \frac{d(C_{Sl})}{dt} = -\mu . C_{Ml} . V . \frac{1}{Y_{MSg}} - r_P . C_{Ml} . V . \frac{1}{Y_{PS}} - r_m . C_{Ml} . V \quad \text{(E - 5.1)}$$

For species M:

$$\frac{Vd(C_{Ml})}{dt} = \mu . C_{Ml} . V \quad \text{(E - 5.2)}$$

For species P:

$$\frac{Vd(C_{Pl})}{dt} = r_P . C_{Ml} . V \quad \text{(E - 5.3)}$$

Figure 5.3 shows a time progression of substrate, cell mass and product mass concentrations in a batch reactor, with consideration of cell decay. The curves in Figure 5.3 were generated using Equations 5.1 - 5.3. After inoculation, the substrate concentration decreases, initially slowly, then faster until all substrate has been consumed. Simultaneously, the biomass concentration increases, initially slowly, then faster until it levels off when substrate becomes depleted. In accordance with the schematic growth curve (Figure 5.2), the curves in Figure 5.3 show a lag phase (approx 0 - 5 hours), a log (-arithmic) phase (approx 6 - 12 hours) and a stationary phase (from approx 14 hours). The parameter values are chosen such that product formation exceeds cell mass production. It will be clear that cell decay is most clearly visible from the point where cell and product formation have ceased, ie starting from time = 14 hours. From this point onward, product mass remains virtually constant while cell mass will start to decrease.

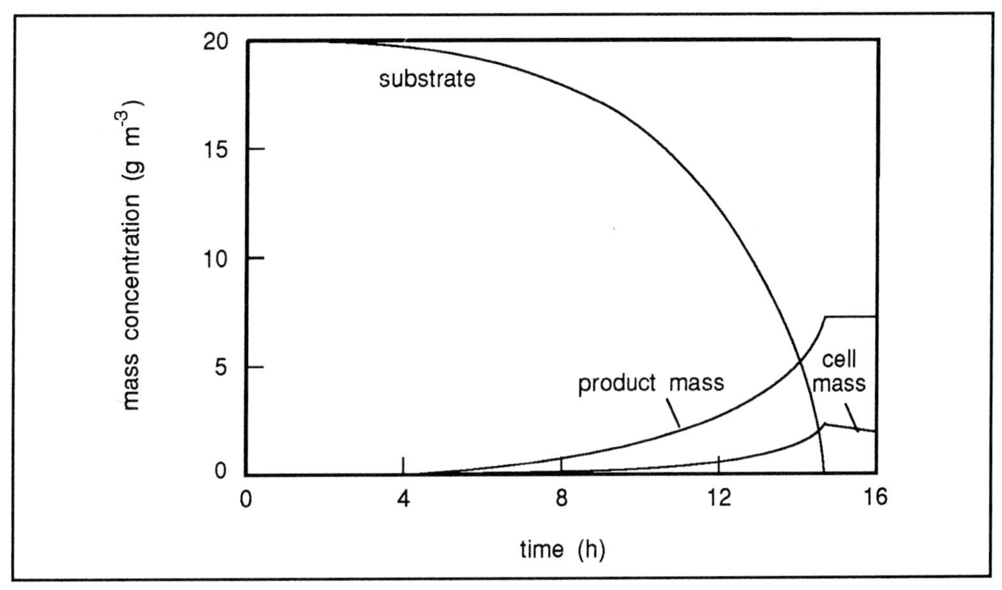

Figure 5.3 Time progression of cell mass, product mass and substrate in a batch process with consideration of cell decay (see text).

5.4 Fed batch reactor

Operating a batch reactor involves a large series of preliminary activities including cleaning, filling with nutrient suspension and, generally, sterilisation of the reactor. After inoculation the bioprocess proceeds through the lag, exponential and stationary phase. The death phase marks the end of the process. The cell mass and/or product are harvested, the reactor is dismantled. To obtain a second batch of cell and/or product mass requires passing through the entire cycle of events again. Thus, the batch reactor goes through a considerable amount of down-time (time in which the reactor is not operational).

Let us do another little calculation. Let us assume that it takes 24h to down-load a batch reactor, clean it, fill it with fresh medium, sterilise it, cool it and inoculate it. Let us also assume that the culture has no lag phase, but immediately enters the log phase on inoculation and has a mean generation time of 1h. Let us also assume that a normal cycle involves 24 generations. We can represent this sequence of events like this:

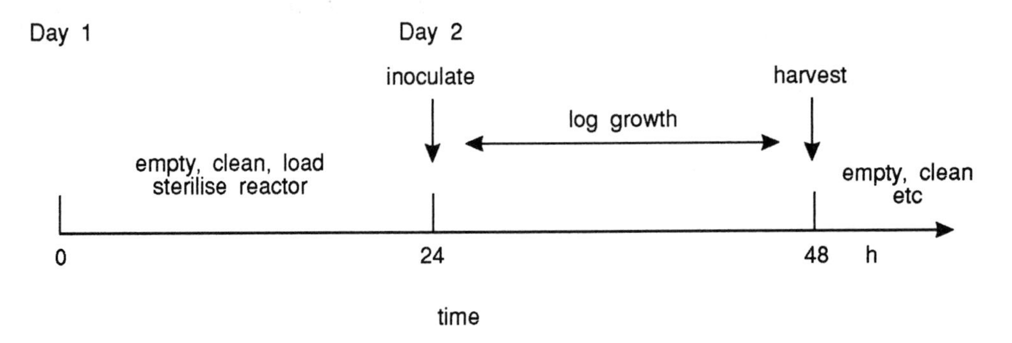

If the product we are interested in is biomass (or a product directly linked to biomass), half of the total quantity which will be produced in the 2 day cycle will be produced in the period 47-48 h[1] (remember the exponential nature of growth $2 \rightarrow 4 \rightarrow 8$ etc). Another way of thinking about this is that the catalyst for producing biomass or products is biomass itself. Thus the reactor is only effectively making biomass (or products) at a high rate when it contains significant quantities of biomass. A feature of a batch system is that for most of the time the reactor is either empty or containing only small quantities of biomass.

In contrast, the CFSTR does not incorporate the need of restarting a new process after completion of the old one (once started CFSTR can operate, theoretically, for indefinite lengths of time and can be run with fairly high biomass concentrations). However, the CFSTR is characteristic for a constant environment and therefore unsuited for bioprocesses in which the idiophase is, time wise, independent of the trophase. Moreover, the CFSTR has the disadvantage of operational complexity (eg sterility). The fed-batch process is developed to meet both the requirement of a more or less continuous operation and of a continuously changing environment (Figure 5.4).

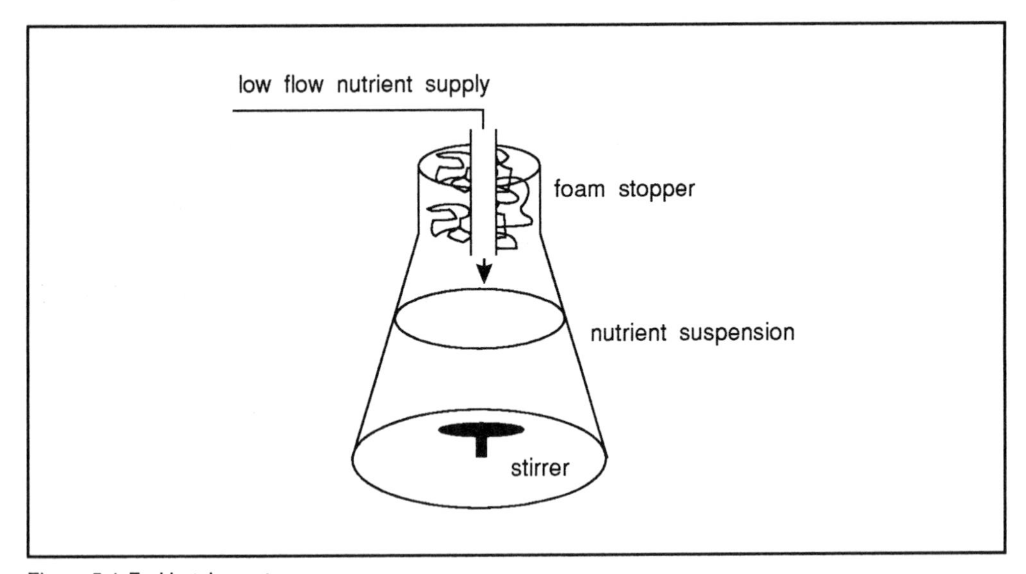

Figure 5.4 Fed batch reactor.

The fed-batch reaction (FBR) is a batch reactor to which, when the nutrients approach depletion, fresh nutrients are added. In other words, the reactor is fed. It is assumed that the concentration of the nutrients added is so high that volume changes are negligible (justifying the batch part of the name).

The profile of the feed rate of the nutrients is generally a function of time, changing Equation 5.1 to:

For species S:

$$\frac{Vd\,(C_{SI})}{dt} = -\,\mu\,.\,C_{MI}\,.\,V\,.\,\frac{1}{Y_{MSg}} - r_P\,.\,C_{MI}\,.\,V\,.\,\frac{1}{Y_{PS}}$$

$$-\,r_m\,.\,C_{MI}\,.\,V + F\,(t) \qquad \text{(E - 5.4)}$$

where F(t) is time dependent function of the influent flow rate.

An example of an industrial fed-batch process is the production of penicillin by *Penicillium chrysogenum*. In this process, the cellular growth rate of the mycelium during the trophophase is high allowing this phase to be short. On the other hand, the cellular growth rate during the idiophase is very low and the emphasis in this phase is on the production of penicillin.

A sanitary engineering example of a batch process is a landfill to which new garbage (nutrients) is added constantly while no material is removed. Or the 'fill and draw' sedimentation tank, a tank that is filled with water, allowed to stand for some time so that suspended material can settle after which the clarified supernatant (top water layer) is 'drawn' off.

SAQ 5.3

Identify which of the following are true and which are false.

1) Batch reactors are most productive over only a short period of the batch cycle.

2) In principle the productivity of a CFSTR is more or less constant over time.

3) A compound which is only produced in the log phase of a batch culture is a primary product.

4) Fed batch reactors are more productive than simple batch reactors.

Summary and objectives

In this brief chapter we have discussed batch reactors. We have described them as closed systems, but have pointed out that in many cases this is not strictly true. Subsequently we described the growth curve of batch cultures and introduced some important terminology used in this area. We then pointed out some advantages of batch reactors (eg operational simplicity, separation of trophophase and idiophase) and disadvantages (eg low productivity). We also explained the advantages of using a fed batch system.

Now that you have completed this chapter you should be able to:

- use the terms extensive properties, intensive properties, trophophase, idiophase, primary product and secondary product appropriately;
- identify primary and secondary products from supplied data;
- list the advantages and disadvantages of batch reactors and CFSTRs.

Plug flow reactor

Plug flow reactor

6.1 Introduction

In this chapter we will examine plug flow reactors (PFR). We begin by describing the general features of these types of reactors and then produce balance equations. This is again a short chapter so you should be able to complete it in one session.

6.2 General features of plug flow reactors

The plug flow reactor (PFR) is characterised by a high length over width (breadth) ratio (Figure 6.1).

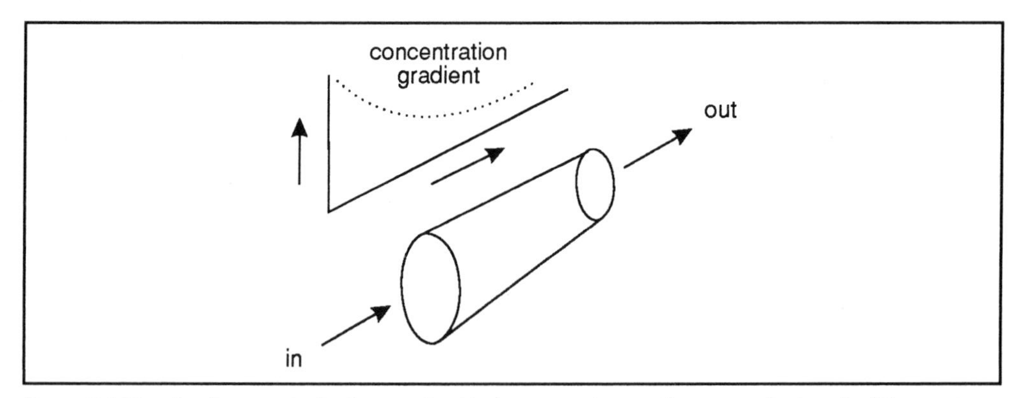

Figure 6.1 The plug flow reactor is characterised by concentration gradients over the length of the reactor.

Liquid mixing is assumed to be limited to the radial direction. The nutrient suspension and microbial inoculum enter the reactor at the inlet, whilst cell mass and microbial product leave through the outlet. Conversion reactions occur between inlet and outlet. Because of the absence of mixing in the direction of flow, the PFR can be regarded as a continuum of batch reactors. The further from the point of inlet, the further the process has advanced. Therefore, the PFR shows a concentration gradient in the direction of flow (Figure 6.1). Because source material enters the reactor continuously and products leave the reactor continuously and, while moving from inlet to outlet, the reaction content passes through an entire series of environmental changes, the PFR can be regarded as a continuous series of batch reactors.

Examples of plug flow reactors are plentiful. A river could be considered as a PFR if no nutrients enter (eg oxygen) or leave (eg CO_2) between the points between which the reactor is defined. Similarly, the oxidation ditch or carrousel, and the sedimentation tank in environmental engineering approach the condition of being PFRs.

SAQ 6.1

Compare the PFR with the BR and CFSTR in terms of mixing, process continuity and reactor environment by completing the following table.

Use terms like: fully; constant; varying; continuous and discontinuous.

	CFSTR	BR	PFR
Mixing			
Process continuity			
Reaction environment			

6.3 Balance equations for the PFR

Consider a PFR with flow rate F, initial substrate concentration C_{Si}, initial cell mass concentration C_{Mi}, volume V and cross-sectional area A (Figure 6.2). Because of the fact that the liquid composition in the PFR is not homogeneous (as opposed to the CFSTR and batch reactors), ie close to the inlet the reactor contents will be primarily substrate. Approaching the outlet, microbial cell and product mass will dominate, applying the conservation of mass rules requires the application of micro-balances.

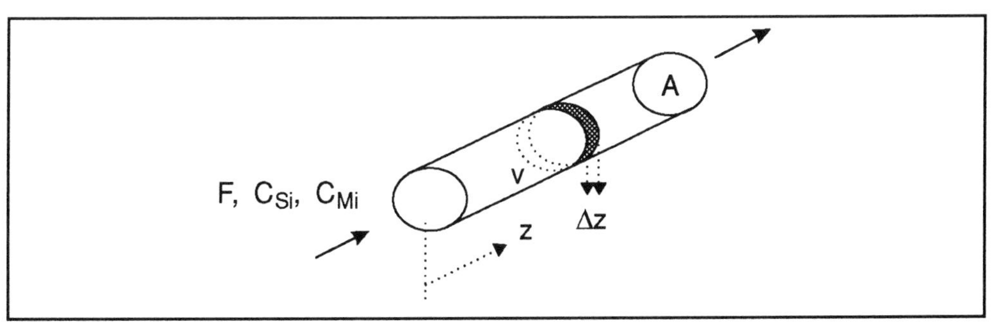

Figure 6.2 Plug flow reactor (see text for meaning of symbols).

For a three species system. S $\rightarrow$ M + P, mass balance equations across a differential element (differential only in the z-direction) in the PFR with volume A . Δz, velocity of flow v_z in the direction of z, yields the following:

For species S:

$$\frac{dC_S}{dt} \cdot A\Delta z = -\left[\left[A \cdot v_z \cdot C_{Sz}\right]_z - \left[A \cdot v_z \cdot C_{Sz}\right]_{z+\Delta z} \right] + R_{Sz} \cdot A \cdot \Delta z$$

(E - 6.1)

For species M:

$$\frac{dC_M}{dt} \cdot A\Delta z = -\left[\left[A \cdot V_z \cdot C_{Mz}\right]_z - \left[A \cdot V_z \cdot C_{Mz}\right]_{z+\Delta z} \right] + R_{Mz} \cdot A \cdot \Delta z$$

(E - 6.2)

For species P:

$$\frac{dC_P}{dt} \cdot A\Delta z = -\left[\left[A \cdot v_z \cdot C_{Pz}\right]_z - \left[A \cdot v_z \cdot C_{Pz}\right]_{z+\Delta z} \right] + R_{Pz} \cdot A \cdot \Delta z$$

(E - 6.3)

where:

z = distance from reactor inlet [L]

Δz = small element of reactor length [L]

A = cross-sectional area of reactor [L^2]

R_{Sz}, R_{Mz}, R_{Pz} = specific rate of respectively substrate consumption, cell mass production and product formation [$M\ L^{-3}\ t^{-1}$]

C_{Sz}, C_{Mz}, C_{Pz} = concentration of substrate, cell mass and product mass at point z [$M\ L^{-3}$].

Assuming steady-state conditions (conditions which do not change in time), dividing Equation 6.1-6.3 by A and Δz and substituting v_z for F/A results in:

For species S:

$$v_z \cdot \frac{\left[C_{Sz}\right]_z - \left[C_{Sz}\right]_{z+\Delta z}}{\Delta z} = R_{Sz} \qquad \text{(E - 6.4)}$$

For species M:

$$v_z \cdot \frac{\left[C_{Mz}\right]_z - \left[C_{Mz}\right]_{z+\Delta z}}{\Delta z} = R_{Mz} \qquad \text{(E - 6.5)}$$

For species P:

$$v_z \cdot \frac{\left[C_{Pz}\right]_z - \left[C_{Pz}\right]_{z+\Delta z}}{\Delta z} = R_{Pz} \qquad \text{(E - 6.6)}$$

where v_z = is the velocity of flow in the z-direction.

v_z of course = flow rate/area = F/A.

Π It would be constructive for you to derive Equations 6.4 - 6.6 yourself. We have done a derivation for Equation 6.4 as an example.

Under steady state conditions:

$$\frac{dC_s}{dt} = 0 \quad \frac{dC_M}{dt} = 0 \text{ and } \frac{dC_P}{dt} = 0$$

Thus for species S:

$$0 = -\Big[\left[A \,.\, v_z \,.\, C_{Sz}\right]_z - \left[A \,.\, v_z \,.\, C_{Sz}\right]_{z+\Delta z} \Big] + R_{Sz} \,.\, A \,.\, \Delta z$$

Divide through by A, then:

$$0 = -\Big[\left[v_z \,.\, C_{Sz}\right]_z - \left[v_z \,.\, C_{Sz}\right]_{z+\Delta z} \Big] + R_{Sz} \,.\, \Delta z$$

$$= v_z \Big[\left[C_{Sz}\right]_z - \left[C_{Sz}\right]_{z+\Delta z} \Big] = R_{Sz} \,.\, \Delta z$$

Thus: $$v_z \,.\, \frac{\left[C_{Sz}\right]_z - \left[C_{Sz}\right]_{z+\Delta z}}{\Delta z} = R_{Sz}$$

Now derive Equation 6.5 and 6.6 for yourself, using this as a model.

Taking in Equations 6.4 - 6.6 the limit of Δz to zero results in:

For species S:

$$v_z \,.\, \frac{dC_{Sz}}{dz} = R_{Sz} \tag{E - 6.7}$$

For species M:

$$v_z \,.\, \frac{dC_{Mz}}{dz} = R_{Mz} \tag{E - 6.8}$$

$$v_z \,.\, \frac{dC_{Pz}}{dz} = R_{Pz} \tag{E - 6.9}$$

with the differential form in Equations 6.7 - 6.9 describing the rate of increase of substrate consumption, microbial cell production and product mass formation. Including the rate expressions similar to those used for the derivation of the balance equations for the CFSTR (Section 4.7), results in:

For species S:

$$v_z \frac{dC_{Sz}}{dz} = -\mu . \frac{C_{Mz}}{Y_{MSo}} - r_P . \frac{C_{Mz}}{Y_{Ps}} \qquad \text{(E - 6.10)}$$

For species M:

$$v_z . \frac{dC_{Mz}}{dz} = \mu . C_{Mz} \qquad \text{(E - 6.11)}$$

For species P:

$$v_z . \frac{dC_{Pz}}{dz} = r_P . C_{Mz} \qquad \text{(E - 6.12)}$$

with: $\mu = \frac{\mu_{max} . C_{Sz}}{K_S + C_{Sz}}$ and $r_P = k_{Pg} . \mu + k_{Pn}$

The boundary conditions for these three differential equations are:

$C_{Sz} = C_{Si}$ for $z = 0$ (E - 6.13)

$C_{Mz} = C_{Mi}$ for $z = 0$ (E - 6.14)

$C_{Pz} = 0$ for $z = 0$ (E - 6.15)

We remind you that μ = specific growth rate (t^{-1})

μ_{max} = maximum specific growth rate (t^{-1})

r_P = specific rate of formation of product (M_s, $L^{-3} t^{-1}$)

Y_{Mso} = observed biomass yield. ($M_m M_s^{-1}$)

Y_{PS} = microbial product yield

C_{MZ} = concentration of biomass at point z along the reactor ($M_m L^{-3}$)

k_{Pg} = growth-associated product formation coefficient ($M_P M_m^{-1}$)

k_{Pn} = non-growth associated product formation coefficient ($M_p M_m^{-1} t^{-1}$)

Using these equations enable us to calculate the concentration of substrate, biomass and product along the reactor providing we know the various coefficients (eg μ, Y_{MSo}, r_P) and flow rate. Alternatively if we measure the concentrations of substrate, biomass and product along the reactor we can determine the coefficients. In principle therefore, we can design a plug flow reactor to achieve a particular reduction in substrate concentration or product yield using the relationships described in Equation 6.10 - 6.12.

Figure 6.3 shows typical progression curves of substrate, cell and product mass concentrations through a PFR. The curves are obtained by numerical integration of Equations 6.10 and 6.12. For the parameter values chosen, substrate is depleted at the end of the reactor length.

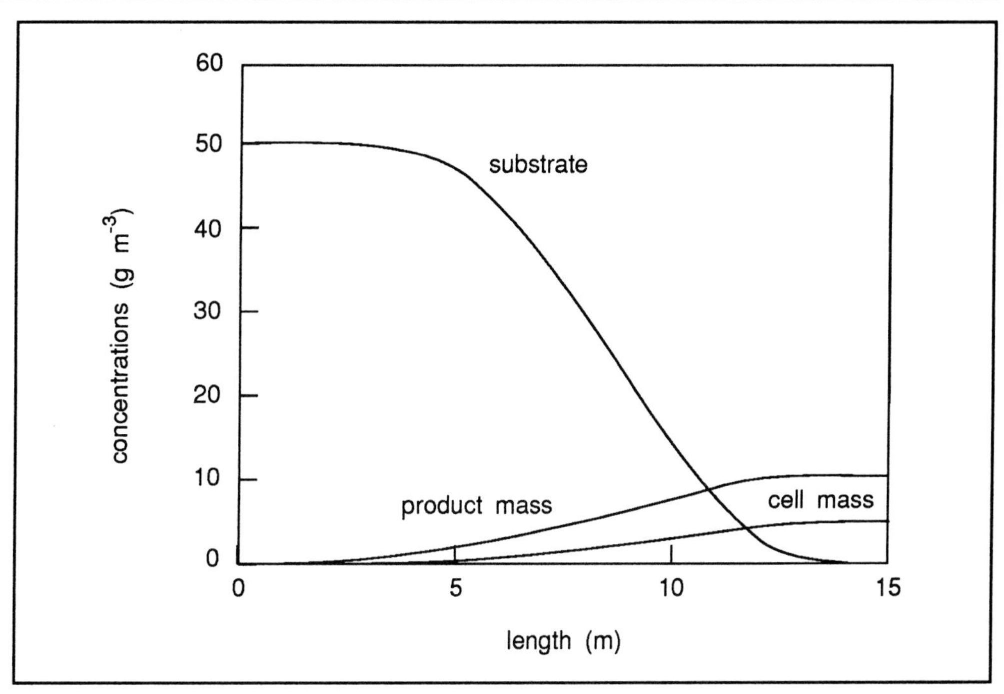

Figure 6.3 Progression of substrate, cell and product mass concentrations in a plug flow reactor.

SAQ 6.2

If a zero substrate concentration is measured at the outlet of a PFR, is it possible to uniquely determine the substrate consumption rate in a PFR? (Give reasons for your answer).

SAQ 6.3

1) Given that the flow rate of a plug flow vessel of cross sectional area 1 m^2 is 0.1 m^3 s^{-1} and that the specific growth rate is 1 h^{-1} and the biomass concentration is 100 g C m^{-3} at a point 10 m along the reactor, what is the rate of change in biomass concentration along the vessel at this point?

2) Using the same conditions, if the specific rate of formation of product is 1 g product C/g biomass C h, what is the rate of change of product concentration at this point?

3) Given that the microbial product yield is 0.1g product C/g substrate C and that the substrate concentration is changing at a rate of 1200g m^{-4} 10m along the reactor, what is the observed biomass yield?

Summary and objectives

In this chapter we examined the general features of plug flow reactors and derived balance equations for substrate consumption, product formation and biomass production. The approach used was to consider the plug flow reactor as a linear series of batch reactors.

Now that you have completed this chapter you should be able to:

- make a comparison between PFR, BR and CFSTR in terms of mixing, process continuity and reactor environment;
- use relationships between velocity of the flow through the reactor, specific product formation rate and yield coefficients to calculate changing substrate, biomass and product concentrations along the reactor from supplied data;
- give examples where PFR-like conditions are employed in practise.

Reactor systems with cell recycling

Reactor systems with cell recycling

7.1 Introduction

The reactor systems discussed perviously, ie the continuous flow stirred tank reactor (CFSTR), the batch reactor (BR) and the plug flow reactor (PFR), all had in common that the reacting species (the biocatalyst, microbial cell) were freely suspended and the reactor liquid was a homogeneous brew (in the PFR homogeneity applies to the radial direction only). In addition, the CFSTR and PFR have in common that transformation (bioreaction) takes place in the time between the material entering and leaving the reactor, ie they are once-through reactors.

Suspended growth systems are the oldest man-made reactors known. They will continue to play an important role in bioprocess engineering. However, certain disadvantages can be listed that have resulted in design alternatives and in the development of radically different reactors. One such alternative is to use cell mass recycling as a means to increase cell mass concentration. We examine such a strategy in this chapter. Cell immobilisation is another alternative and this will be discussed in Chapter 8.

In this chapter we will first describe the general principles behind cell recycling including reasons why this might be desirable. We will then set up and apply balance equations for CFSTRs fitted with recycling.

7.2 General principles of cell recycling

external biomass feedback

Recycling implies that a certain fraction of a process stream from some point in the process is separated and introduced in a point more upstream in the same process (Figure 7.1).

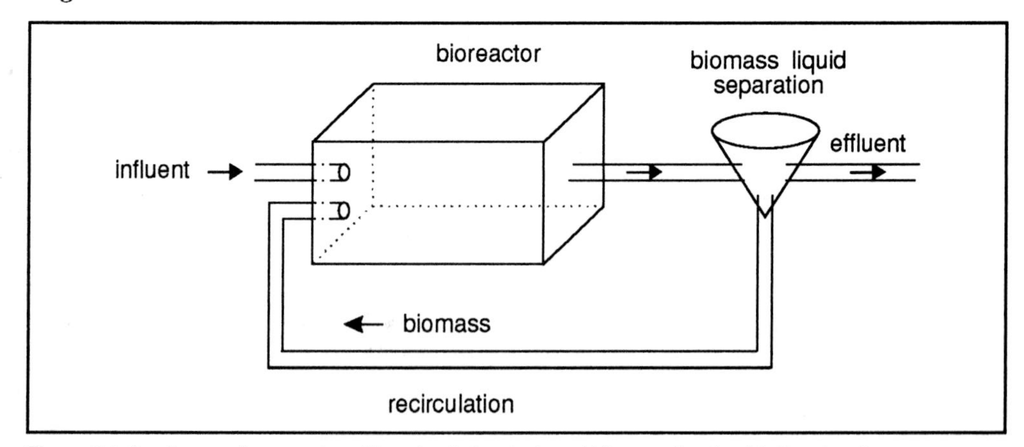

Figure 7.1 Continuous flow reactor with recirculation (external biomass feedback). See text for a description.

We might describe the system shown in Figure 7.1 as a CFSTR with an external biomass feedback loop. This is not however the only configuration. Figure 7.2 illustrates a different configuration. In this the effluent is removed in two streams. One removes culture, the other removes effluent through a filter which effectively removes some (or all) of the biomass from the effluent. Such a system could be described as a CFSTR with internal biomass feedback. Whether or not an internal or external biomass feedback loop is used, the effect is to increase the amount of biomass in the reactor.

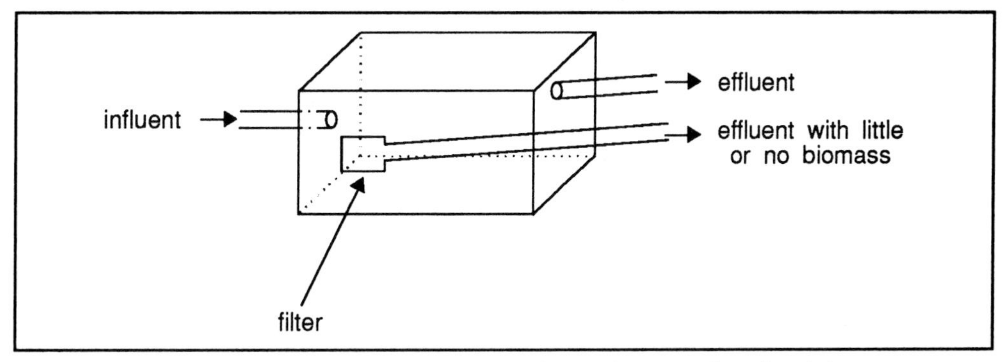

Figure 7.2 Continuous flow reactor with internal biomass feedback (see text for a description).

Another type of reactor which has internal biomass feedback is one in which the reactor has two chambers (Figure 7.3). In this, biomass growth occurs in the lower chamber. In the upper chamber, cells settle out and fall back into the growth chamber. Again two effluent streams are collected: the one from the upper chamber has a low biomass concentration; the other effluent stream leaving from the lower chamber has a high biomass concentration. Although this looks quite different to that illustrated in Figure 7.2, the net effect is the same: biomass is concentrated in the reactor.

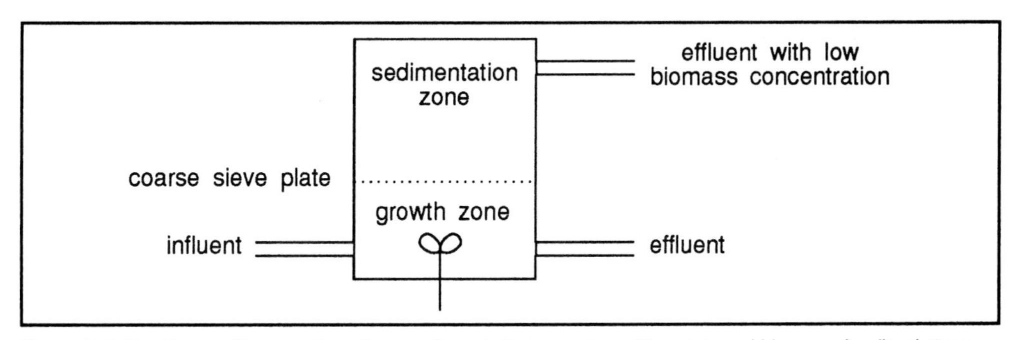

Figure 7.3 Continuous flow reactor with a sedimentation zone to achieve internal biomass feedback (see text for a description).

In this chapter we will focus onto CFSTR with external biomass feedback as illustrated in Figure 7.1. You should note that we adopt the common practice of using the terms recycling and recirculation interchangeably.

7.3 Reasons for using cell recycling

Various reasons for applying recycling can be described as:

- reuse of effluent components. The liquid flow leaving a once-through reactor may still contain unused substrate but certainly will contain (re-)usable microbial cells;
- increasing cell mass concentration. The maximum cell concentration in a once-through reactor is determined by cell kinetic and stoichiometric coefficients, the influent substrate concentration and the dilution rate. By recycling part of the effluent flow, this cell mass concentration can be increased. Thus we will have more biocatalysts in the reactor;
- dilution of reactor influent. Adding a recycle flow to the reactor influent line helps to reduce the influent concentration.

It may be clear that different objectives will demand different fractions of the effluent to be recycled, eg a liquid fraction, a cell mass fraction or a combination thereof. The mass balance approach for a CFSTR with recycling step is discussed in the next section.

7.4 Balance equations for a CFSTR with recycling

When determining mass balance equations for reactor systems consisting of more than one reactor component, the first step is to define the boundaries between which the balance equations are to be determined. For example, consider a system consisting of two reactor components, one for a biological conversion process, the other for a physical separation process (see Figure 7.4). This figure looks quite complex so let us explain it in a little more detail.

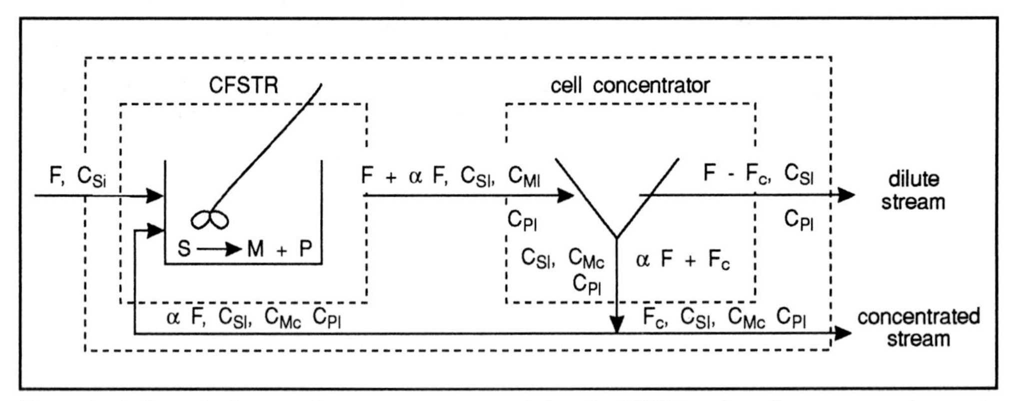

Figure 7.4 Schematic diagram of a reactor system consisting of a CFSTR and a cell concentrator (see text for description).

First let us follow the flow through the system. Begin with the CFSTR. The biological process takes place in a CFSTR. The effluent from the CFSTR overflows into a concentrator for the physical separation of microbial cells from liquid. The concentrator effluent is discharged. A fraction of the concentrate is also discharged while the remainder of the fraction is recycled (this configuration is reminiscent of the activated sludge-/sedimentation tank - combination in waste water treatment). Balance equations can be determined around reactor and concentrator separately or around the two units combined. The principles involved in setting up mass balance equations for reactor systems are similar to those for reactors only.

Now let us tackle the symbols on Figure 7.4

Π As you read through this section, look at Figure 7.4 and note where the symbols occur. Alternatively draw out the structure of the system as shown below and then write on the symbols where they should be. You can then check your product with that illustrated in Figure 7.4. This is one way you can ensure that you have understood the system.

The following figure shows the outline of a CFSTR with a cell concentrate:

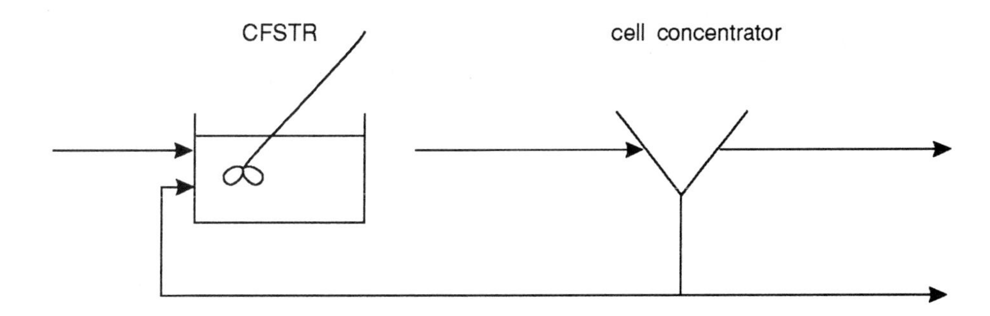

Consider a CFSTR-concentrator combination that is fed with substrate for the continuous production of cell and product mass. Note that we assume that the cell concentrator does not remove substrate or product from the effluent from the CFSTR. Thus the concentration of substrate entering the cell concentrator (C_{SI}) = the concentration of the substrate in the out flow of the cell separator. Note also we have assumed that the cell concentrator removes all the biomass, thus in the dilute stream C_M = 0. The following flows into and out of the CFSTR and concentrator can be distinguished.

Reactor (biological process):

influent:	substrate C_{Si} flow rate F
recycle in:	substrate C_{Sl} flow rate αF (α is the fraction of F that is recycled into reactor) cell mass C_{Mc} product mass C_{Pl}
effluent:	substrate C_{Sl} flow rate $F + \alpha F$ cell mass C_{Ml} product mass C_{Pl}

Concentrator (physical process):

influent:	substrate C_{Sl} flow rate $F + \alpha F$ cell mass C_{Ml} product mass C_{Pl}
concentrate:	substrate C_{Sl} concentrated flow rate F_c concentrated cell mass C_{Mc} product mass C_{Pl}
effluent (dilute stream):	substrate C_{Sl} flow rate $F\text{-}F_c$ product C_{Pl}

Balance equations for a CFSTR-separator (Figure 7.4) are as follows:

Reactor - for substrate species:

verbally:

change in substrate in the reactor	=	flow of substrate into the reactor	-	flow of substrate out of the reactor	-	substrate used for growth	-	substrate used for product formation

Note that we have simplified this by including cell decay and maintenance in the term 'substrate used for growth'. Numerically:

$$V_R \cdot \frac{dC_{Sl}}{dt} = F \cdot C_{Si} + \alpha F \cdot C_{Sl} - (1+\alpha) \cdot F \cdot C_{Sl} - \mu \cdot \frac{C_{Ml}}{Y_{MSo}} \cdot V - r_P \cdot \frac{C_{Ml}}{Y_{PS}} \cdot V \qquad \text{(E - 7.1)}$$

where V_R is the volume of the reactor

Reactor - for cell mass species:

verbally:

change in biomass in the reactor	=	flow of biomass into the reactor	-	flow of biomass out of the reactor	+	growth of biomass

Numerically:

$$V_R \, . \, \frac{dC_{Ml}}{dt} = \alpha F \, . \, C_{Mc} - (1 + \alpha) \, . \, F \, . \, C_{Ml} + \mu \, . \, C_{Ml} \, . \, V \qquad \text{(E - 7.2)}$$

Reactor - for product mass species:

Verbally:

change in product in the reactor	=	flow of product into the reactor	-	flow of product out of the reactor	+	production of product in the reactor

Numerically:

$$V_R \, . \, \frac{dC_{PL}}{dt} = \alpha F \, . \, C_{Pl} - (1 + \alpha) \, . \, F \, . \, C_{Pl} + r_P \, . \, C_{Ml} \, . \, V \qquad \text{(E - 7.3)}$$

in which: μ = specific microbial growth rate (t^{-1}), r_P = specific cellular product formation rate ($M_P \, M_M^{-1} \, t^{-1}$)

We can use an analogous approach for the concentrator. For simplicity we have omitted the verbal equation. Perhaps you would like to write these in under the equations. We have done the first one for you.

Concentrator - for substrate species:

$$V_c \, . \, \frac{dC_{Sl}}{dt} = C_{Sl} \, . \, (1 + \alpha) F - (F_c + \alpha F) \, . \, C_{Sl} - (F - F_c) \, . \, C_{Sl} \qquad \text{(E - 7.4)}$$

change of substrate in the concentrator	=	flow of substrate into the concentrator	-	flow of substrate in the biomass concentrate stream leaving the concentrator	-	flow of substrate in the dilute biomass stream leaving the concentrator

Concentrator - for cell mass species:

$$V_c \cdot \frac{dC_M}{dt} = C_{Ml} \cdot (1 + \alpha) F - (F_c + \alpha F) \cdot C_{Mc} \quad \text{(E - 7.5)}$$

Concentrator - for product mass species:

$$V_c \cdot \frac{dC_{Pl}}{dt} = C_{Pl} \cdot (1 + \alpha) F - (F_c + \alpha F) \cdot C_{Pl} - (F - F_c) \cdot C_{Pl} \quad \text{(E - 7.6)}$$

F_c = flow rate of concentrated cell mass leaving concentrator $[L^3 t^{-1}]$

F = flow rate into the reactor $[L^3 t^{-1}]$

C_{Mc} = concentration of concentrated cell mass $[M_M L^{-3}]$

C_{Ml} = concentration of biomass in the influent

C_{Sl} = concentration of substrate $[M_S L^{-3}]$

C_{Pl} = concentration of the product $(M_P L^{-3})$

V_c = volume of concentrator $[L^3]$

α = recycle fraction of F

Note that in contrast to the equations for product mass and substrate, in terms of cell mass the concentrator can not be regarded as completely mixed, ie the concentration of cell mass depends upon the location in the concentrator. Therefore, the cell mass concentration in the accumulation term, dC_M/dt (Equation 7.5), is not provided with the subscript 'l'.

The Equations 7.1 - 7.6 look quite complex, yet if we approach these logically, by writing first a verbal equation and then converting the different verbal components into a numerical form, they become relatively straightforward.

We can however simplify them further, especially under steady state conditions.

Remember that under steady state conditions that $V \cdot \frac{dC_{Sl}}{dt}$, $V \cdot \frac{dC_{Ml}}{dt}$ and $V \cdot \frac{dC_{Pl}}{dt}$ all become 0. Remember also that dilution rate $D = \frac{F}{V}$

Thus from Equation 7.1 we can write:

$$D \cdot \frac{C_{Si} - C_{Sl}}{C_{Ml}} = \frac{\mu}{Y_{MSo}} + \frac{r_P}{Y_{PS}} \quad \text{(E - 7.7)}$$

since:

$$V \frac{dC_{Sl}}{dt} = F C_{Si} + \alpha F C_{Sl} - (1 + \alpha) F C_{Sl} + - \mu \cdot \frac{C_{Ml}}{Y_{MSo}} V - r_P \cdot \frac{C_{Ml}}{Y_{PS}} \cdot V$$

(Equation 7.1)

Thus at steady state:

$$0 = FC_{Si} + \alpha F C_{Sl} - (1+\alpha) F C_{Sl} - \mu \frac{C_{Ml}}{Y_{MSo}} V - r_P \frac{C_{Ml}}{Y_{PS}} V$$

$$0 = FC_{Si} + \alpha F C_{Sl} - F C_{Sl} - \alpha F C_{Sl} - \mu \frac{C_{Ml}}{Y_{MSo}} V - r_P \frac{C_{Ml}}{Y_{PS}} V$$

$$0 = F C_{Si} - F C_{Sl} - \mu \frac{C_{Ml}}{Y_{MSo}} - r_P \frac{C_{Ml}}{Y_{PS}} V$$

$$F (C_{Si} - C_{Sl}) = \mu \frac{C_{Ml}}{Y_{MSo}} V + r_P \frac{C_{Ml}}{Y_{PS}} V$$

Dividing through by V gives:

$$D (C_{Si} - C_{Sl}) = \mu \frac{C_{Ml}}{Y_{MSo}} + r_P \frac{C_{Ml}}{Y_{PS}}$$

$$\text{Thus } D \frac{C_{Si} - C_{Sl}}{C_{Ml}} = \frac{\mu}{Y_{MSo}} + \frac{r_P}{Y_{Ps}}$$

Equation 7.7 was derived earlier (Equation 4.23) for a continuous flow reactor. Apparently, microbial growth and product formation rate in a reactor system with recycling are independent from the size of the recycled fraction.

Assuming steady-state conditions for Equation 7.2 gives:

$$D \, . \left[1 + \alpha - \alpha \, . \, \beta \right] = \mu \qquad \text{(E - 7.8)}$$

with

$$\beta = \frac{C_{Mc}}{C_{Ml}} \qquad \text{(E - 7.9)}$$

β is called the biomass concentration factor (in contemporary biotechnology) or the sludge thickening factor (in wastewater engineering). This is perhaps obvious since it measures the ratio of the concentration of biomass leaving the concentration to the concentration in the concentrate leaving the reactor.

Equation 7.8 indicates that in a reactor system with recycling, the microbial growth rate is not necessarily equal to the dilution rate (recycling is 'numerically defined' for α unequal to zero and β unequal to 1). For $\beta = C_{Mc}/C_{Ml} = 1$, meaning no concentration of microbial cells, the effect of recycling on microbial growth rate is zero.

The microbial growth rate μ has to be at least zero. Therefore, from Equation 7.8 it follows that:

$$\left[1 + \alpha - \alpha \, . \, \beta \right] \geq 0 \qquad \text{(E - 7.10)}$$

and thus:

$$\alpha \leq \left[\frac{1}{\beta - 1} \right] \qquad \text{(E - 7.11)}$$

From the relationship between α and β found in Equation 7.11 (Figure 7.5) it can be seen that when β increases, α can be reduced for a similar cell mass loading rate into the reactor. Although possibly trivial, this is an important finding from the point of view of operational cost.

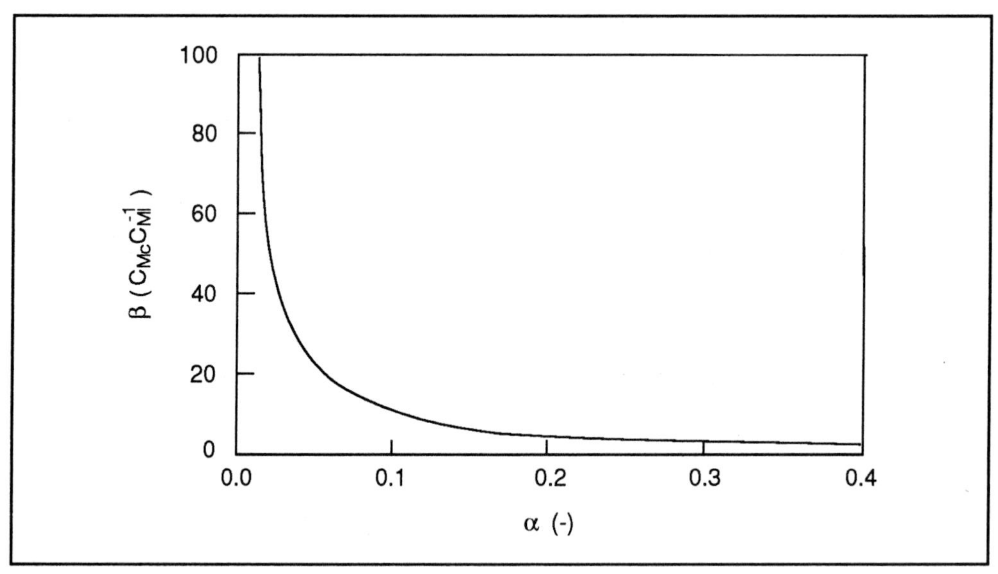

Figure 7.5 Relationship between α (recycled fraction) and β (biomass concentration factor) to give a constant biomass loading (see text for details).

We have covered a lot of ground so far so let us stop and do some calculations.

SAQ 7.1

If the substrate concentration entering a CFSTR fitted with recycling is 500 g substrate C m^{-3}, its concentration leaving the vessel is 10 g substrate C m^{-3} and the dilution rate is 2 h^{-1} and the growth rate is 1 h^{-1}, what is the observed biomass growth yield? Assume no other products are made and that the biomass concentration is 500 g biomass C m^{-3}.

SAQ 7.2

1) What will the concentration factor of a CFSTR with recycling be if the dilution rate is 2 h^{-1}, specific growth rate of the organism is 1.2 h^{-1} and the fraction of the flow returning to the bioreactor is 40%?

2) If the biomass concentration in the concentrate flowing back into the reactor is 500 g biomass C m^{-3}, what is the concentration of biomass in the reactor?

SAQ 7.3

A CFSTR fitted with cell recycling has a concentration factor of 4 and 20% of the flow is recycled. What is the specific growth rate of the culture in this vessel when the dilution rate is 2 h^{-1}?

The assumption of steady-state conditions for Equations 7.3, 7.4 and 7.6 does not result in new information. However, assuming steady-state conditions for Equation 7.5, division by C_{Ml} and F, and rearranging results in:

$$\left[1 + \alpha - \alpha \frac{C_{Mc}}{C_{Ml}}\right] = \frac{F_c}{F} \cdot \frac{C_{Mc}}{C_{Ml}} = (1 + \alpha - \alpha\beta) = \frac{F_c}{F} \cdot \frac{C_{Mc}}{C_{Ml}} \qquad \text{(E - 7.12)}$$

Substituting Equation 7.8 ($D[1 + \alpha - \alpha\beta] = \mu$) in Equation 7.12 results in:

$$\mu = \frac{F_c}{F} \cdot \frac{C_{Mc}}{C_{Ml}} \cdot D \qquad \text{(E - 7.13)}$$

Substitution of F/V for D changes Equation 7.13 into:

$$\mu = \frac{F_c}{V} \cdot \frac{C_{Mc}}{C_{Ml}} \qquad \text{(E - 7.14)}$$

Let us see if we can put into words what this equation states. Equation 7.14 describes the microbial growth rate μ in a reactor system employing recycling of concentrated biomass. The product $F_c C_{Mc}$ is called the cell mass production rate [$M_M t^{-1}$] in the system. The product $V C_{Ml}$ equals the total cell mass in the reactor. Therefore, dividing the cell mass production rate by the cell mass in the system equals the specific cell mass production rate, which is the specific cellular growth rate and commonly indicated with the symbol μ.

The inverse of the microbial growth rate, ie θ_b, is called the mean cell residence time ($\theta_b = 1/\mu$). The inverse of the dilution rate, ie θ_h, is the hydraulic residence time, ($\theta_h = 1/D$). The mean cell residence time (sludge age) is a function of the hydraulic residence time according to Equation 7.15.

$$\theta_b = \frac{F}{F_c} \cdot \frac{C_{Ml}}{C_{Mc}} \cdot \theta_h \qquad \text{(E - 7.15)}$$

Π Perhaps you would like to prove to yourself that this equation holds. Begin with Equation 7.14 or Equation 7.13 and substitute in $\theta_b = \frac{1}{\mu}$ and $\theta_h = \frac{1}{D} = \frac{V}{F}$.

When C_{Mc} equals C_{Ml}, no concentration of cells takes place and hence, F_c equals F. In this situation, the mean biomass residence time θ_b and, consequently, microbial growth rate is determined by the dilution rate D (which is, as could be expected from the fact that the concentrator does not work, the situation described before for the CFSTR).

Substituting V/F for θ_h in Equation 7.15 (the inverse of the manipulation preceding Equation 7.14) results in an expression for the sludge age, an important process parameter in wastewater treatment:

$$\theta_b = \frac{V}{F_c} \cdot \frac{C_{Ml}}{C_{Mc}} \qquad \text{(E - 7.16)}$$

In this section we have introduced rather a lot of terms, without looking through the previous section of text, attempt SAQ 7.4.

SAQ 7.4

Match each of the phrases below with the appropriate mathematical relationship.

cell mass production rate	$\frac{V}{F}$
specific cell mass production rate	$\frac{1}{\mu}$
sludge age	$V\,C_{MI}$
mean cell residence time	$\frac{F_c}{V}\,\frac{C_{Mc}}{C_{Ml}}$
hydraulic residence time	$F_c\,C_{Mc}$
concentration factor	$\frac{C_{Mc}}{C_{Ml}}$
total cell mass	$\frac{F}{F_c} \cdot \frac{C_{Mi}}{C_{Mc}} \cdot \theta_h$

The effect of α and β on steady-state substrate and biomass concentrations in a CFSTR concentrator system as a function of dilution rate is shown in Figure 7.6. In this graph, the concentration factor (β) is chosen to be fixed ($\beta = 4$) while the recycled fraction (α) is varied. The effects of varying α for a fixed β are similar to the effects of varying β for a fixed α.

It can be concluded that for a specific concentration factor of the concentrator, varying the recycled fraction can be used to fine tune the cell mass concentration and hence the substrate concentration in the reactor (generally, the concentration factor is less easy to manipulate than the recycled fraction).

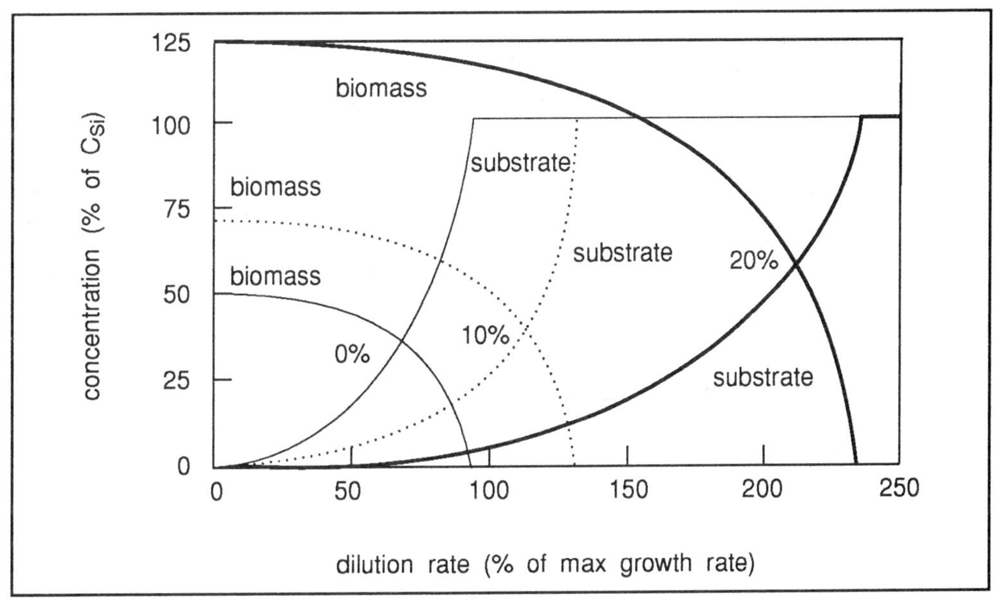

Figure 7.6 The effect of recycling on cell mass and substrate concentration in the reactor. Numbers refer to % recycled fraction.

Note that in systems in which cells are recycled we can use dilution rates higher than the maximum specific growth rate (μ_{max}) and still get significant conversion of the substrate. It also means that the productivity of the system (dilution rate x biomass concentration) can be very much greater than systems without cell recycling.

Π In Figure 7.6, at what dilution rate do we get wash out when 20% of the outflow is recycled.

The answer is approximately at about 230% of μ_{max}.

Π What % of the of the substrate is used when the dilution rate is 150% of μ_{max} and 20% of the outflow is recycled?

The answer is about 85% since only 15% of the substrate remains.

Π Let us complete this chapter with a fairly tough exercise.

Set up the mass balance equations for CFSTR with recycle whereby excess cell mass removal takes place from the CFSTR. Compare the results with those from the system discussed previously in which the excess cell mass is removed from the concentrator. Use the following figure to help you. Do not worry if you cannot completely solve it, for we have provided a solution below. It would however be good practice to really try this before looking at our solution.

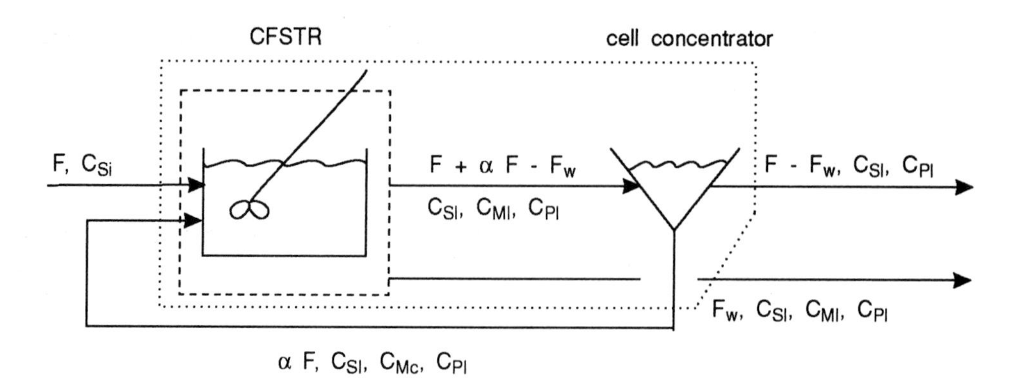

Begin by writing verbal equations for substrate, biomass and product. We will give you a start.

For the reactor the substrate balance is:

change in substrate	=	flow of substrate into the reactor	-	flow of substrate out of the reactor (2 flows)	-	substrate used for growth	-	substrate used for product formation

Use a similar approach for biomass and product for both the reactor and the concentrator.

Here is our solution. We have omitted writing in the individual verbal equations.

Reactor - for substrate species:

$$V \cdot \frac{dC_{Sl}}{dt} = F \cdot C_{Si} + \alpha F \cdot C_{Sl} - (F + \alpha F - F_w) \cdot F \cdot C_{Sl} - F_w \cdot C_{Sl} - \mu \cdot \frac{C_{Ml}}{Y_{MSo}} \cdot V - r_P \cdot \frac{C_{Ml}}{Y_{PS}} \cdot V \quad \text{(E - 7.17)}$$

Reactor - for cell mass species:

$$V \cdot \frac{dC_{Ml}}{dt} = \alpha F \cdot C_{Mc} - (F + \alpha F - F_w) \cdot F \cdot C_{Ml} - F_w \cdot C_{Ml} + \mu \cdot C_{Ml} \cdot V \quad \text{(E - 7.18)}$$

Reactor - for product mass species:

$$V \cdot \frac{dC_{Pl}}{dt} = \alpha F \cdot C_{Pl} - (F + \alpha F - F_w) \cdot F \cdot C_{Pl} - F_w \cdot C_{Pl} + r_P \cdot C_{Ml} \cdot V \quad \text{(E - 7.19)}$$

Assuming steady-state conditions (ie $V \frac{dC_{Sl}}{dt} = 0$) results, after some rearrangements, from Equation 17:

$$D \cdot \frac{C_{Si} - C_{Sl}}{C_{Ml}} = \frac{\mu}{Y_{MSo}} + \frac{r_P}{Y_{PS}} \qquad \text{(E - 7.20)}$$

Equation 7.20 is identical to Equation 4.23, derived for a CFSTR without recycle, or to Equation 7.7, derived for a CFSTR with recycle and cell mass discharge from a concentrator.

Equation 7.21 gives:

$$\mu = D \cdot \left(1 + \alpha - \alpha \cdot \frac{C_{Mc}}{C_{Ml}}\right) \qquad \text{(E - 7.21)}$$

Equation 7.21 is identical to Equation 7.8. Moreover, rearranging Equation 7.19 results in:

$$D \cdot C_{Pl} = r_P \cdot C_{Ml} \qquad \text{(E - 7.22)}$$

which was found earlier (Equation 4.25). Consequently, in a mathematical sense, there is no difference whether a CFSTR-concentrator combination discharges cell mass from the concentrator or from the CFSTR.

7.5 Concluding remarks

In Section 7.3, we gave the reasons for using CFSTRs with cell recycling. We would like to re-emphasise them here. CFSTRs fitted with cell recycling means that we can have much higher biomass concentrations in the reactor. Since the rate of removal of substrate or formation of product is related in some way to the amount of catalyst present, it should be self-evident that CFSTRs equiped with recycling have much greater bioconversion capabilities per unit volume of vessel than do conventional CFSTRs. Thus we can either produce much more product or remove much more substrate than could be achieved with simple CFSTRs, or we can use a much smaller vessel to achieve the same amount of bioconversion. Either way there is clearly economic advantage in using such a system. A good contrast would be to consider the amount of biocatalyst in a CFSTR with cell recycling with that of a batch reactor. We made the case that with a batch reactor the reactor was virtually empty of catalyst for much of the cycling time (see Chapter 5). In contrast, a continuous flow reactor fitted with recycling maintains a high rate of bioconversion continuously (except of course if there is mechanical breakdown or contamination).

The main problem with CFSTRs with cell recycling is that they are prone to microbial contamination. This is not a particular problem where mixed cultures of opportunistic organisms are used (eg in sludge digestion) but is critical when pure cultures are required (eg in the production of pharmaceuticals). It is not surprising therefore that these types of reactors find particular use in such processes as waste water treatment.

One final point to make is that since $\mu = D(1 + \alpha - \alpha\beta)$, Equation 7.8, then such systems can be operated at much higher dilution rates. If we make the approximation that the critical dilution rate $D_c = \mu_{max}$ for a simple CFSTR, then we can use a much higher dilution rate (D) before we reach a value equivalent to μ_{max}. Thus with a biomass feedback system, we can have a much faster flow through the vessel and effectively higher productivity.

Summary and objectives

In this chapter we have described continuous flow stirred tank reactors with cell recycling. We have explained the reasons behind using recycling and established balance equations for such systems.

Now that you have completed this chapter you should be able to:

- list reasons for using CFSTRs with cell recycling;
- write balance equations for CFSTRs with cell recycling;
- calculate a variety of parameters relating to CFSTRs with recycling from supplied data using relationships derived from balance equations;
- use appropriately a wide variety of terms used in describing CFSTRs with recycling including cell mass production rate, specific cell mass production rate, mean cell residence time and concentration factor;
- write down numerical expressions for a wide variety of terms applied to CFSTRs with recycling.

Attached growth reactors

Attached growth reactors

8.1 Introduction

In the previous chapters we have examined reactor systems in which the biocatalyst (cells) is present in suspension. We now turn our attention to reactors in which the cells are attached to a solid. In the first part of the chapter we will make some general points about attached growth reactors and introduce some of the terms applied to these systems. We will then discuss the differences between immobilised cell and cell suspension systems and briefly consider immobilised enzyme systems. We will then briefly list the various techniques employed in enzyme and cell immobilisation and describe some technological cell immobilisation systems. Finally processes involving natural cell immobilisation are discussed together with the modelling of the gradients in biofilms and the mass balance equations for natural biofilm systems.

8.2 Attached growth reactors and microbial ecology

Attached growth (fixed film, biofilm, immobilised cell) reactors are reactors in which the biocatalyst is attached to (adsorbed onto) a solid or suspended substratum. Other reaction species may be present both in suspension and in the attached phase. Attachment of the biocatalyst is a means by which the cell mass is retained in the reactor.

There are a number of terms that are commonly used in attached growth systems:

- the term 'substratum' is used to indicate that the biocatalyst is sorbed to something with a density significantly different from that of water or cell material;

- microbial slime layers are referred to as immobilised cell layers 'fixed films' and 'biofilms'. The first term is used mainly in the realm of biotechnology, the second primarily in public health engineering. The third term appears most neutral. In this text, the three terms are used interchangeably.

Although virtually all microbial organisms have the potential to adsorb onto a substratum, technological application of cell sorption processes is only recently being made use of on a wider scale. The most well-known examples are in wastewater engineering where trickling filters and, more recently, rotating biological contactors and fluidised bed reactors are being used. Animal cell culture systems also make use of attached cell reactors.

Let us take a brief excursion into the area of microbial ecology to demonstrate that attached growth reactors are not only man-made. Microbial activity can be found in all natural waters. In the larger rivers passing through large population centres (eg river Rhine, Nile, Mississippi), micro-organisms abound in the liquid phase (water) because of an ample food supply (high concentrations of dissolved degradable organics in the water). At the other end of the spectrum, in the pristine mountain stream with very low concentrations of organics, the number of micro-organisms in the liquid phase is relatively limited, more being present adsorbed onto boulders and other hard surfaces submerged in the water.

Again different situations may arise when the water contains large concentrations of suspended material due to erosion (a wet season phenomenon). Although these waters have generally a high concentration of organisms in the water, a high proportion of organisms is usually adsorbed onto the particulate material. This too can be regarded as an attached growth system.

Let us now consider why attached (immobilised) cell systems are attractive to biotechnologists.

8.3 Immobilisation or suspension?

If practically speaking all micro-organisms exhibit the capacity to attach to a substratum, why then does attachment play a role in the trickling filter and other attached growth reactors in wastewater engineering but not in the (suspended growth) reactor systems discussed in earlier chapters?

The key to this question is dilution rate. All systems discussed previously are operated at a dilution rate $D \leq D_c$ (critical dilution rate, the rate at which cells can complete at least one cell division before being washed out). Under those conditions, the cells remain for a sufficient length of time in the reactor. Consequently, being attached does not offer an advantage in terms of availability and use of substrate.

On the other hand, at a dilution rate above the wash-out rate, only cells that have attached remain in the reactor. Suspended cells are being washed out. Therefore, immobilised cell reactors can be operated at a dilution rate above the wash-out rate.

This important finding can be examplified by comparing the cell mass productivity of attached and suspended growth systems. The rate of substrate consumption (proportional to the rate of cell mass production) in a suspended growth reactor is determined by the availability of substrate and by the duration of the cells in the reactor. Therefore, a plot of substrate consumption rate against dilution rate shows a gradual rise in cell mass production rate until a maximum is reached, followed by a drop to zero at the critical dilution rate.

Π Draw a stylised graph of this and then compare it with Figure 4.5.

With attached growth systems, this gradual rise (showing the improvement of the availability of substrate) with increasing dilution rate also occurs. But once substrate is optimally available, cell mass production rate remains more-or-less constant in spite of an increase in dilution rate (Figure 8.1).

The productivity against dilution rate plot for a mixed attached and suspended growth systems is also shown in Figure 8.1. The curve for suspended growth is similar to the one shown in an earlier chapter (Figure 4.5). The curve for attached growth resembles a saturation model. The sum of the two curves is indicated for such, ie a mixed attached and suspended cell mass system is shown.

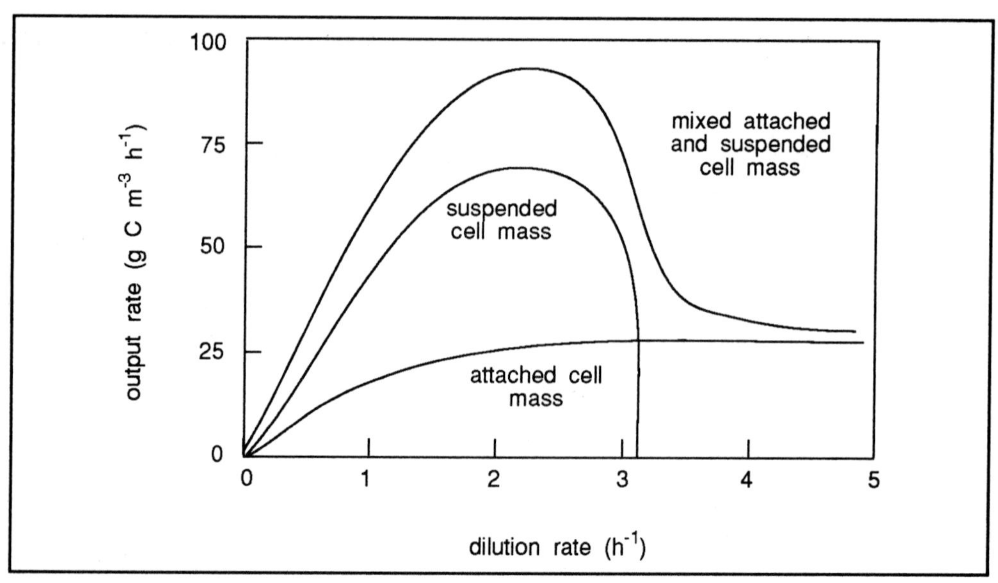

Figure 8.1 Models showing cell mass production rate against dilution rate for suspended or/and attached growth systems.

It should be noted that for a mixed attached and suspended growth process, there will never be a sharp change in cell mass productivity as shown in Figure 8.1. Rather, the curve will follow the dotted line shown in Figure 8.1.

8.4 Differences between attached and suspended growth systems

Π Before reading on see if you can write down a list of some of the advantages of using attached growth systems. Then compare your list with the one below.

In comparing attached and suspended growth systems a few differences become obvious. In attached growth processes:

- systems can be operated at dilution rates exceeding the critical dilution rate;
- the cell mass per unit volume can be considerably higher than the cell mass per unit volume in a suspended growth system;
- cell mass and product mass productivity can be considerable higher than in suspended growth processes because we can use greater densities of cell mass per unit volume;
- cells are (primarily) in the solid phase (sorbed) while substrate and microbial product are in the liquid phase. This intrinsic separation reduces the need for a separation step in the down stream processing of the reactor effluent;
- the biocatalyst is continuously reused. In suspended growth systems, reuse can only be implemented by recirculating the biocatalyst.

Your list might have been a little different to ours, but we hope you included the main points.

Π What do you consider is the most serious disadvantages of the attached growth reactor?

Again you may have thought of several. The main problem we might identify is that if the layer of attached cells was thick then the rate at which substrate might reach the inner most cells might become rate limiting thus this part of the biomass would be ineffective (ie substrate limited). It may of course even decay.

8.5 Comparison of immobilised cells and enzymes

cost of purified enzymes

The words 'attached growth' as used here, refer both to attached microbial cells and to attached enzymes. In the early days of immobilisation, interest focused almost exclusively on the use of enzymes. Enzymes, the components governing all processes in the microbial cell, are used in purified form to perform a single, or sometimes a multiple, enzymatic conversion. However, the purification of enzymes is generally costly limiting the use of enzymes.

cells as dilute catalysts

Cells, in comparison, are relatively dilute catalysts, by which we mean that, relative to the volume of the cell, the enzyme concentration is low. Cells contain several non-enzymic components which can contaminate the product. In addition, cells contain numerous other enzymes which may compete for the substrate or may even metabolise the product further. Yet, it can be said that immobilised cells represent a highly flexible catalyst system due to the many enzyme activities they contain. Therefore, cells have gradually become an attractive alternative to enzymes in applications of immobilised biocatalysts. If you would like to read further on this aspect we recommend the review of Birnbaum S, Lasson P O and Mosbach K (1986), 'Immobilised biocatalysts, the choice between enzymes and cells' (In Webb C, Black G M and Atkinson B. 'Process engineering aspects of immobilised cell systems'. 'The Institution of Chemical Engineers' pp 35-53).

SAQ 8.1

Assign each of the following either to immobilised cell systems or to immobilised enzyme systems.

1) The cost of preparing the biocatalyst is usually high.

2) The system may contain activities which compete for the substrate.

3) Is most likely to lead to a purer product.

4) Is able to transform substrate through many intermediates to make products that are chemically quite different to the substrate.

5) The biocatalyst is likely to be present in low concentrations.

6) Substrate access to the biocatalyst may be restricted.

8.6 Kinds of immobilisation

Although it was stated earlier that virtually all micro-organisms have the potential to adsorb to a substratum, it may be desirable to help the formation of the biofilm. Reasons for this are to obtain an immobilised layer of for example a specified thickness, predefined composition and with specific properties, or just simply to reduce the time needed to obtain an actively functioning layer. The various methods of biocatalyst immobilisation are shown in Figure 8.2.

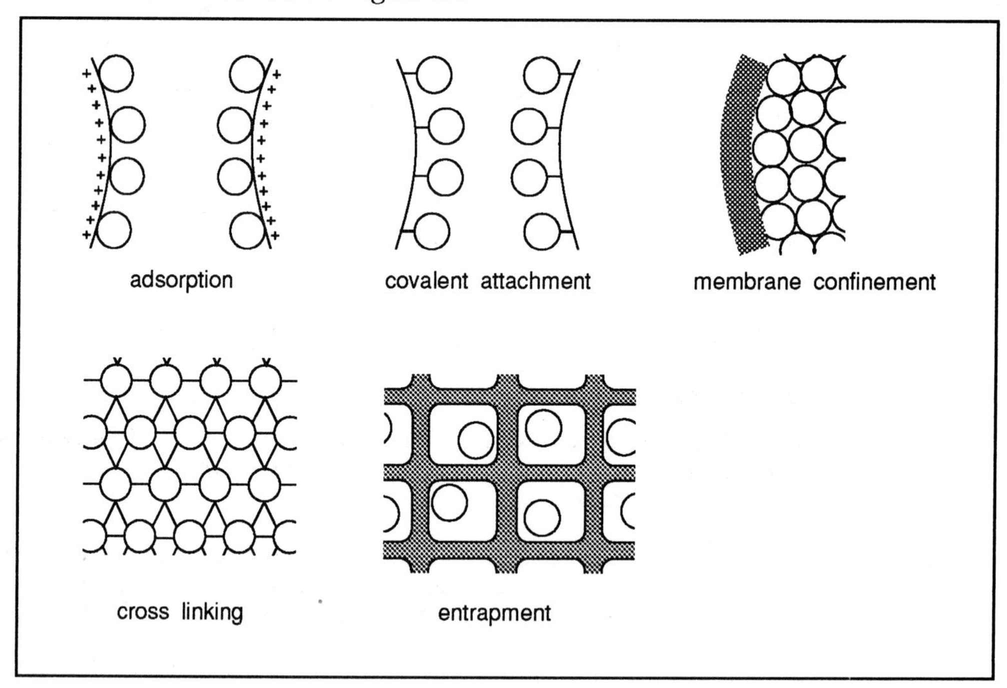

Figure 8.2 Various kinds of immobilisation.

chemical and physical methods of immobilisation

We do not propose to elaborate on the various methods of immobilisation here. This topic is discussed in the BIOTOL text 'Technological Applications of Biocatalysts'. You should however realise that the methods make use of ionic binding (including van der Waal's forces), illustrated in Figure 8.2 as adsorption; chemical binding, illustrated in Figure 8.2 as covalent attachment and cross linking; entrapment in a gel (eg alginate) or confinement by a membrane. We can therefore identify both chemical and physical methods of immobilisation.

Immobilised cells can be used in a variety of ways (Figure 8.3). Notice that the immobilised cells can be viable or non-viable. They can be permeabilised and a single enzyme or multiple enzymes within the cells may be used. The following discussion will concentrate on biofilms following spontaneous immobilisation. These biofilms have the advantage of *in situ* proliferation and, hence, *in situ* synthesis of specific enzymes.

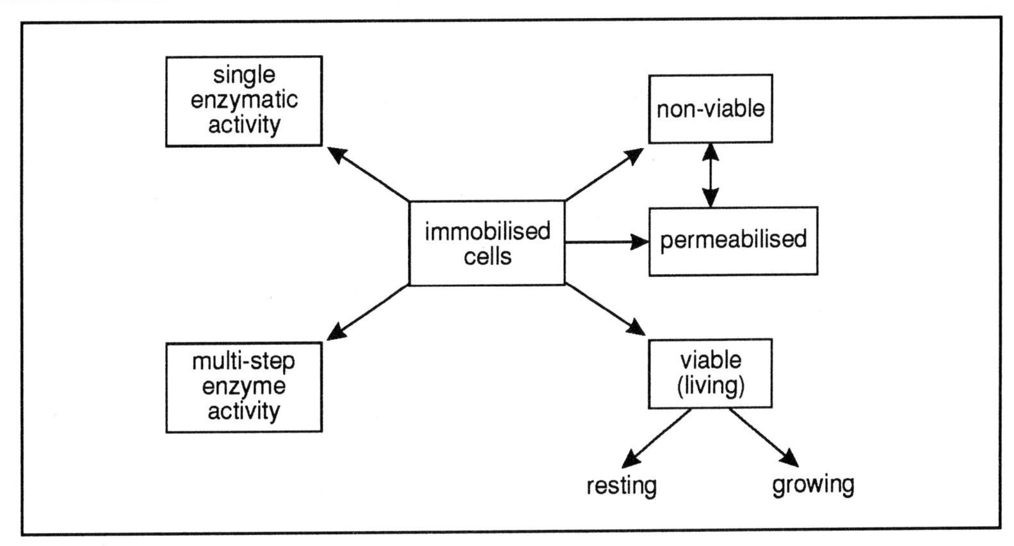

Figure 8.3 The various uses of microbial cells.

8.7 Cell immobilisation systems in practice

8.7.1 Types of systems

Immobilised cell systems come in many different forms. Without attempting to cover all of the possibilities we give a brief description of the immobilised systems used in wastewater engineering.

- trickling filters, packed bed filters - wastewater is distributed over a rock-like material covered by a microbial layer. Nutrients contained in the wastewater are adsorbed by the micro-organism on the passage through the filter bed;
- rotating biological contactors - RBC's are formed by disc like sheets partly submerged in wastewater on a horizontal shaft rotating. The discs are covered by a microbial slime layer onto which nutrients are adsorbed from the wastewater;
- fluidised beds - upflow reactors are reactors to which small particulate material is added onto which microbial organisms adsorb. The upflow velocity keeps the particulate material in suspension;
- aerated submerged fixed film reactors - the air phase of the natural aeration, as in RBC treatment, is replaced by artificially supplied air;
- cell encapsulation systems - microbial cells are, during the particulate production phase, encapsulated in beads through which nutrients have to diffuse to, and products diffuse away from, the cell. Primarily in experimental use.

8.7.2 Processes of spontaneous cell immobilisation

Spontaneous cell adsorption and subsequent biofilm accumulation is the result of microbial, chemical and physical processes occurring in the liquid phase, within the biofilm and at the substratum (Figure 8.4). Here we provide a description of the major processes identified in biofilm accumulation. It should be noted that the older the biofilm, the more likely it will be that the processes will occur simultaneously.

Π We will be using quite a lot of descriptive terms as we go through this description. You may find it helpful to write these down on a sheet of paper and write out exactly what you think they mean. We will give you a check list at the end of this section.

Refer to Figure 8.4 as you read the following section, which gives an account of the processes involved in spontaneous cell immobilisation. Note that on this figure we have indicated 6 phases. We will deal with each in turn.

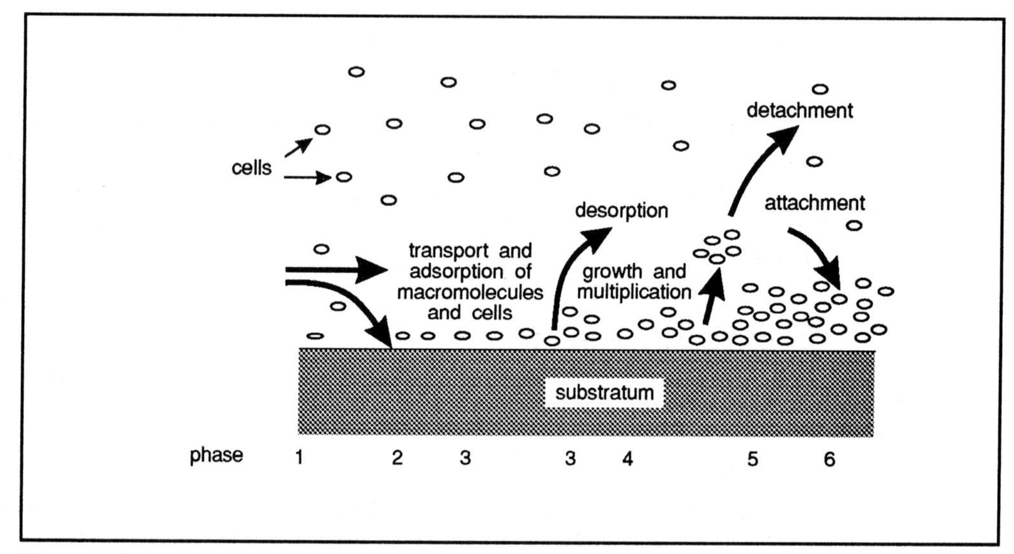

Figure 8.4 Process of biofilm accumulation. Numbers refer to explanations in the text.

Phase 1: Transport and adsorption of organic macromolecules and nutrients

conditioning film

Most natural and process waters contain organic macromolecules and nutrients. These compounds are homogenously distributed in the liquid phase while a fraction will adsorb to solid substrata due to various physical (ie hydrodynamic) and chemical processes. Adsorption of organic macromolecules and nutrients generates a concentration gradient which results in transport of species from the liquid phase to the substratum forming a so called conditioning film, the precise role of which is still unclear. Adsorption of macromolecules and nutrients will continue until the substratum is saturated.

Phase 2: Transport of micro-organisms

transport of cells to substrator

Micro-organisms (dimensions of 0.5 to 10 μm) are transported from the liquid phase to the wetted substratum through a variety of mechanisms: (Brownian) diffusion, gravity, taxis and fluid dynamic forces such as inertia, lift, drag, drainage and downsweep. Motility can contribute significantly to the rate of transport of a microbial cell to the substratum. Transport of microbial species to the substratum will continue until the substratum is saturated.

Phase 3: Adsorption of micro-organisms

reversible adsorption

Micro-organisms adsorb onto the substratum in a reversible or irreversible fashion. With reversible adsorption, adsorbed microbial cells leave the substratum (the cells desorb) before bonding is sufficient to resist the hydrodynamic shear forces.

irreverisble adsorption

Irreversible adsorption occurs following the production of extracellular fibres of sufficient strength, bonding the microbial cell to the substratum.

Phase 4: Reactions within the biofilm

development of concentration gradients

Once organisms are adsorbed, energy is expended for growth, replication, the formation of extracellular products (eg polysaccharides, proteins, small molecules, peptides, antimicrobial agents), and cell maintenance. Microbial activity implies the consumption of microbial nutrients and development of a nutrient concentration difference (gradient) between the biofilm and the liquid phase. This results in diffusion of nutrient species into the biofilm.

Phase 5: Separation of biofilm and associated products

detachment

erosion

sloughing

Parts of the biofilm may separate and become re-entrained in the bulk liquid. When cells are separating from other cells, this process is termed detachment. When individual cells separate from the biofilm this is called erosion while sloughing generally refers to separation of pieces of biofilm (cells and products). In other words erosion and sloughing are two different forms of detachment.

Phase 6: Attachment of cells

attachment

Once a biofilm is established, cells in the liquid phase may still be deposited on or in the biofilm, a process called attachment.

Reactions between the biofilm and the substratum

Reactions may occur between microbial organisms and their products and the substratum. Also reactions may occur between areas of the substratum covered with biofilm and areas not covered. The adsorbed cells grow and reproduce, forming colonies.

glossary of terms

Glossary of special terms used in the description of the processes involved in spontaneous cell immobilisation:

- adsorption - cells sticking directly to the uncolonised substratum;
- attachment - cells sticking to other cells in a biofilm, attached cells do not really 'touch' the substratum directly;
- conditioning film - surface of substratum after initial adsorption of macromolecules and nutrients;
- desorption - cells loosening directly from the substratum;
- detachment - cells loosening from the biofilm, called erosion when it concerns individual cells, sloughing when cells come off as lumps;
- erosion - detachment of individual cells from the biofilm;
- sloughing - detachment of clumps of cells from a biofilm;

- taxis - responsive movement of organisms towards or away from an external stimulus, usually along a chemical (chemotaxis) or light (phototaxis) gradient.

SAQ 8.2

Let us now test your technical vocabulary. Match up the words with the definitions without looking at the glossary given above.

Words	Definitions
attachment	cells sticking to uncolonised substratum
sloughing	detachment of single cells from a biofilm
substratum	cells possessing a mechanism of locomotion
motility	cells sticking to other cells
taxis	detachment of clumps of cells from a biofilm
detachment	cells move up a gradient of an attracting chemical
desorption	a surface onto which cells stick
adsorption	cells loosening from a substratum
erosion	cells loosening from a biofilm

8.7.3 Gradients in biofilms

Transport of mass is the result of gradients. We mean by gradients, differences in the concentration of the mass over a certain distance. In biological processes, gradients often develop as a result of nutrient consumption by micro-organisms.

concentration gradients in suspended systems

In the liquid phase (in suspension), the micro-organisms, with a density approximately equal to the density of water, move virtually at the same velocity and in the same direction as the liquid that surrounds the organism. Therefore, it appears as if the micro-organism 'eats' a hole in the substrate (Figure 8.5a). The closer to the organism, the more the substrate concentration differs from the concentration in the bulk liquid. Consequently, even though the bulk liquid substrate concentration would allow the organism to grow at its maximum growth rate, the actual growth rate of the organism may be significantly less.

The purpose of mixing in a biological reactor, both in suspended and in attached growth processes, is to reduce the distance over which the substrate concentration around the microbial cell differs from the bulk liquid concentration. With substrate concentration determining microbial productivity, it will be clear that the quality of mixing is of prime importance in the operation of biological processes.

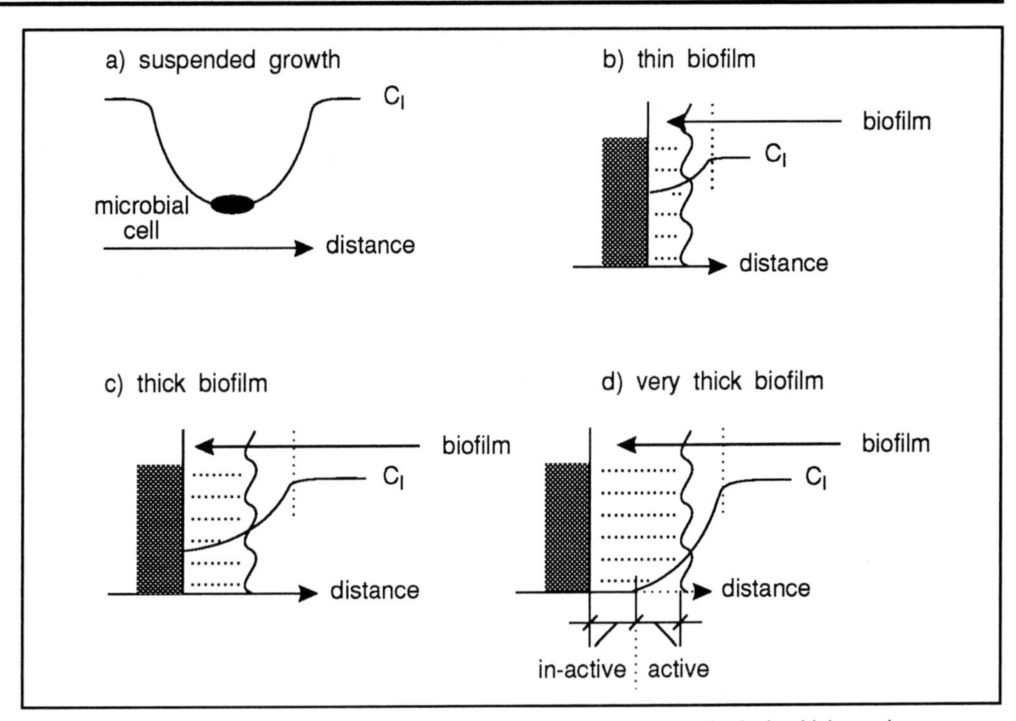

Figure 8.5 a) Concentration gradients for suspended and b), c) and d) attached microbial growth conditions. See text for details. C_l = concentration of substrate in the liquid.

concentration profiles in biofilms

The situation with attached organisms is quite different from that with suspended organisms. Attached organisms are associated with a substratum with a density that is significantly different from that of the organism. Therefore, velocity differences can exist between the bulk liquid in the reactor and the attached microbial cells in the upper layer of a biofilm. In a thin biofilm, cells are surrounded by continuously different media and, hence, by continuously new substrate. The substrate concentration in the upper layer of a biofilm or in a thin biofilm varies little from the bulk liquid substrate concentration (Figure 8.5b).

However, this is not true for the organisms below the top layer in a thicker biofilm. Because of substrate consumption, the deeper in the biofilm the less substrate remains resulting in a distinct concentration gradient (Figure 8.5c). In fact, this reduction of substrate can continue until all substrate is consumed. At that point the bulk of the biofilm becomes inactive (Figure 8.5d).

8.7.4 Diffusion into biofilms

Diffusive transport of matter can result only from a concentration gradient. Consumption of substrate in the biofilm results in a concentration gradient which is the prerequisite for a diffusive flux of substrate into the biofilm. In order to quantify this flux and, therefore, to quantify the profile of the gradient of substrate in the biofilm, a microbalance across an infinitesimal thin layer of biofilm, perpendicular to the direction of the flux, can be set up (Figure 8.6).

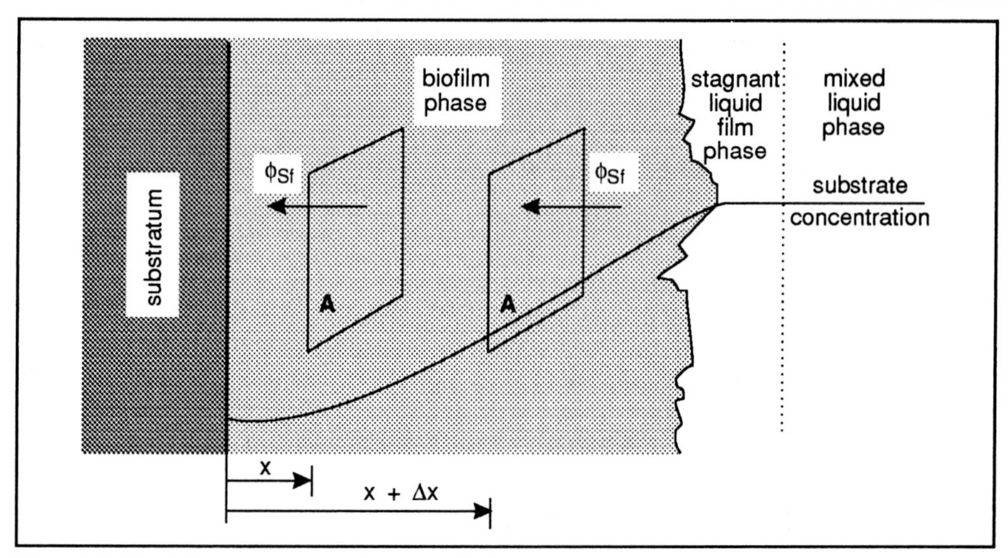

Figure 8.6 Microbalance approach for flux into a biofilm (see text for details).

Consider the flux of substrate S into the biofilm, ϕ_{Sf}. This flux passes through surface area A in a direction opposite to the direction of the ordinate. Considering that the convective flux in a biofilm is negligible, the mass balance for a volume element Δx . A can be set up:

change in substrate in volume A . Δx	=	flow into the volume	-	flow out of the volume	-	amount used by growth

Thus numerically:

$$\frac{dC_{Sf}}{dt} \cdot \Delta x \cdot A = \left[- A \cdot \phi_{Sf} \Big|_{x+\Delta x} - \left(- A \cdot \phi_{Sf} \right) \Big|_{x} \right] - \mu \cdot \frac{C_{Mf}}{Y_{MSo}} \cdot \Delta x \cdot A \qquad \text{(E - 8.1)}$$

where:

C_{Sf} = mass concentration of substrate in the biofilm ($M_S\ L^{-3}$)

C_{Mf} = mass concentration of microbial cells in the biofilm ($M_M\ L^{-3}$)

ϕ_{Sf} = flux of substrate mass into biofilm ($MS\ L^{-2}\ t^{-1}$)

Y_{MSo} = observed microbial cell mass produced per mass of substrate consumed ($M_M M_S^{-1}$)

Although from a cell physiological point of view, Y_{MSo} is the observed cell yield in suspension, in practice cell yields in the biofilm and in suspension are taken to be the same.

Dividing by Δx and A and taking the limit of Δx going to zero allows us to write this difference equation as a differential equation:

$$\frac{dC_{Sf}}{dt} = -\frac{d\phi_{Sf}}{dx} - \mu \cdot \frac{C_{Mf}}{Y_{MSo}} \qquad \text{(E - 8.2)}$$

The flux term, ϕ_{Sf} can be written as:

$$\phi_{Sf} = - D_{Sf} \cdot \frac{dC_{Sf}}{dx} \qquad \text{(E - 8.3)}$$

where D_{Sf} = mass diffusion coefficient of substrate S into biofilm f ($L^2 t^{-1}$).

Substitution of Equation 8.3 into Equation 8.2 results in:

$$\frac{dC_{Sf}}{dt} = D_{Sf} \cdot \frac{d^2C_{Sf}}{dx^2} - \mu \cdot \frac{C_{Mf}}{Y_{MSo}} \qquad \text{(E - 8.4)}$$

Assuming an overall biofilm thickness of L_f, two boundary conditions for this differential form can be listed:

at $x = L_f$, the concentration of substrate in the biofilm equals the substrate concentration in the bulk liquid.

[Note however that according to the theory that presumes the presence of a film layer of laminar flow conditions between the biofilm and the bulk liquid, the substrate concentration at the interface between biofilm and film layer will in practice be less than the substrate concentration in the bulk liquid. This is due to the low rate of diffusion of substrate molecules through this film layer as a result of which this film layer forms a relatively significant diffusional resistance.]

at $x = 0$, the substrate consumption rate is zero resulting in a zero substrate gradient. In mathematical terms:

$$\frac{dC_{Sf}}{dx} = 0$$

This will remain 0 at all times, thus Equation 8.4 can be rewritten:

$$\frac{d^2C_{Sf}}{dx^2} - \mu \cdot \frac{C_{Mf}}{Y_{MSo}\, D_{Sf}} = 0 \qquad \text{(E - 8.5a)}$$

Written in another way:

$$\frac{d^2C_{Sf}}{dx^2} = \mu \frac{C_{Mf}}{Y_{MSo}\, D_{Sf}} \qquad \text{(E - 8.5b)}$$

In other words the concentration gradient is related to the specific growth rate, the biomass concentration, the observed cell growth yield and the diffusion coefficient.

Equation 8.5b gives the parabolic nature of the substrate concentration gradient in a biofilm, consistent with the curve shown in Figure 8.5.

Similar forms can be derived for the concentration of product mass formed in the biofilm. In contrast, microbial cell mass only leaves the biofilm by detachment and, hence, no concentration gradient of cell mass can be derived.

SAQ 8.3

1) Does a faster growth rate of biomass increase or decrease the concentration gradient of substrate in a biofilm?

2) Does a higher biomass concentration increase or decrease the concentration gradient of substrate in a biofilm?

3) Does a higher diffusion coefficient increase or decrease the concentration gradient in a biofilm?

4) Does a higher cell growth yield increase or decrease the concentration gradient in a biofilm?

8.7.5 Balance equations for attached growth systems

In the discussion of balance equations for suspended microbial growth, attention was focused on the description of some of the relevant reactor configurations. Balance equations for attached growth systems are often described independent of the reactor configuration, ie focus is on the species exchange between the liquid and the attached phase. If species exchange can be modelled, than the specific details of the reactor configuration can be added separately.

processes considered in attached growth reactors

The processes considered in the attached growth reactor are similar to those for the suspended growth reactors. These are that nutrients are converted into microbial cell mass and microbial product mass. Therefore, mass balance equations are considered for a two phase - three species system. The phases are: liquid (indicated with a subscript 'l'), attached or film phase (indicated with a subscript 'f'). The species are: nutrients, cells and products.

For the sake of completeness it should be noted that microbial substrate conversion also concerns a third phase, the gas phase. Substrate is converted to microbial cells, cellular products and CO_2. We will not however consider the gas phase here. Consideration of the gas phase is included in the BIOTOL text 'Operational Modes of Bioreactors'.

Before setting up mass balances, the space within which mass is to be balanced has to be determined (Figure 8.7).

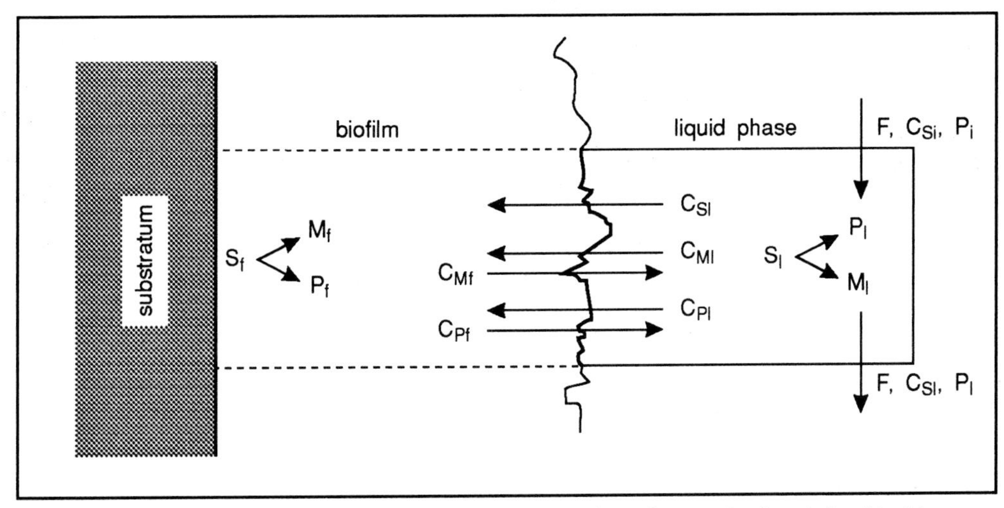

Figure 8.7 Boundaries defining the space for the balance equations. See text for description. M = biomass, P = product, S = substrate, C_S = substrate concentration, C_M = biomass concentration, C_P = product concentration, F = flow rate, Subscript 'l' = liquid, 'f' = film and 'i' = input.

Two spaces are defined in Figure 8.7, a biofilm space and a liquid space. Nutrients, liquid phase cells and liquid phase product may diffuse from the liquid space into the biofilm space. Nutrients in the biofilm are converted into biofilm cell mass and biofilm product mass. On the other hand, biofilm cells and product may diffuse from the biofilm into the liquid.

For setting up the balance equations, the boundary lines enclose a part of the liquid phase and the adjacent section of the biofilm. Mass transport between the delimited liquid phase space and the bulk liquid is considered, in addition to mass transfer between the delimited spaces of the biofilm and the liquid phase. Balance equations will be discussed for the liquid phase and the biofilm phase for microbial cells (M), microbial product (P) and substrate (S).

Π Before we move on, let us make certain you have understood Figure 8.7. Here we provide a drawing of a biofilm and its associated liquid phase. Without looking at Figure 8.7, place the items listed below on the figure and connect them appropriately by arrows. You can then check your answer with Figure 8.7. If you do this successfully you have clearly understood the processes taking place in the biofilm and are ready to write balance equations.

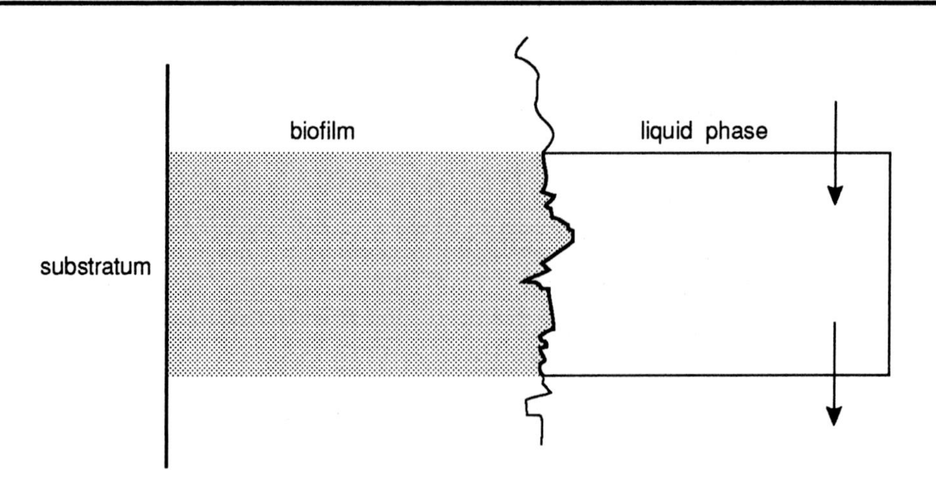

Items to label:

Flow of liquid into and out of the boundary of the liquid phase (F)

Concentration of substrate in the incoming liquid phase (C_{Si})

Concentration of substrate fluxing out of the liquid phase (C_{Sl})

Movement of biomass from the biofilm (C_{Mf})

Movement of biomass into the film (C_{Ml})

Conversion of substrate into product and biomass in the film (S_f, P_f, M_f)

Conversion of substrate into product and biomass in the liquid (S_l, P_l, M_l)

Movement of product from the film into the liquid (C_{Pf})

Movement of biomass from the liquid into the film (C_{Pl})

Balance equations for liquid phase substrate

Verbally we can write:

rate of substrate accumulation per unit volume	=	net rate of substrate inflow per unit volume	-	rate of substrate uptake by biofilm per unit volume	-	rate of substrate uptake in liquid phase for cells and product per unit volume

In numerical form:

$$\frac{dC_{Sl}}{dt} = D\,(C_{Si} - C_{Sl}) - C_{Mf} \cdot \frac{A}{V} \cdot r_{Sf} - C_{Ml} \cdot \left[\frac{\mu_l}{Y_{MSo}} + \frac{r_{Pl}}{Y_{PS}} \right] \qquad \text{(E - 8.6)}$$

where:

C_{Sl} = substrate carbon concentration in liquid phase ($M_S\ L^{-3}$)
t = time (t)
D = dilution rate (t^{-1})
C_{Si} = substrate carbon concentration in influent ($M_S\ L^{-3}$)
C_{Mf} = mass of cell carbon per unit surface are in biofilm ($M_M\ L^{-2}$)
A = substratum area (L^2)
V = reactor volume (L^3)
r_{Sf} = biofilm specific substrate uptake rate ($M_S\ M_M^{-1}\ t^{-1}$)
C_{Ml} = cell carbon concentration in liquid phase ($M_M\ L^{-3}$)
μ_l = specific cellular growth rate in liquid phase (t^{-1})
Y_{MSo} = yield of cell carbon from substrate carbon ($M_M\ M_S^{-1}$)
r_{Pl} = specific product formation rate in liquid phase ($M_P\ M_M\ t^{-1}$)
Y_{PS} = yield of product carbon from substrate carbon ($M_P\ M_S^{-1}$)

Balance equations for liquid phase cell mass

net rate of cell mass accumulation	=	net rate of cell mass inflow	+	rate of cell mass detachment	+	rate of cell growth in liquid phase

$$\frac{dC_{Ml}}{dt} = D\,(C_{Mi} - C_{Ml}) + C_{Mf} \cdot \frac{A}{V} \cdot r_{Md} + C_{Ml} \cdot \mu_l \qquad (E-8.7)$$

where:

C_{Mi} = cell carbon concentration in influent ($M_M\ L^{-3}$)
r_{Md} = specific cell detachment rate (t^{-1})

Balance equations for liquid phase product mass

net rate of product accumulation	=	net rate of product inflow	+	rate of product detachment	+	rate of product formation in liquid phase

$$\frac{dC_{Pl}}{dt} = D\,(C_{Pi} - C_{Pl}) + C_{Pf} \cdot \frac{A}{V} \cdot r_{Pd} + C_{Ml} \cdot r_{Pl} \qquad (E-8.8)$$

where:

C_{Pl} = product carbon concentration in liquid phase ($M_P\ L^{-3}$)

C_{Pi} = product carbon concentration in influent ($M_P\ L^{-3}$)

C_{Pf} = mass of product carbon per unit area of substratum ($M_P\ L^{-2}$)

r_{Pd}= specific rate of product detachment (t^{-1})

Thus we can write balance equations for substrate, cell mass and product mass for the liquid phase. By analogy we can produce similar equations for the attached phase (ie the biofilm).

Balance equations for attached phase substrate

net rate of substrate accumulation	=	net rate of substrate uptake by biofilm	-	rate of substrate uptake for cell growth and product formation

$$\frac{dC_{Sf}}{dt} = C_{Mf} \cdot r_{Sf} - C_{Mf} \cdot \left[\frac{\mu_f}{Y_{MSo}} + \frac{r_{Pf}}{Y_{PS}} \right] \qquad \text{(E - 8.9)}$$

where:

C_{Sf} = mass of substrate carbon per unit area of substratum ($M_S\ L^{-2}$)

μ_f= specific cellular growth rate in biofilm phase (t^{-1})

Note that we have assumed that no substrate moves from the biofilm into the liquid phase.

Balance equations for attached phase cell mass

net rate of cell mass accumulation	= -	net rate of cell mass detachment	+	rate of biofilm cellular growth

$$\frac{dC_{Mf}}{dt} = - C_{Mf} \cdot r_{Md} + C_{Mf} \cdot \mu_f \qquad \text{(E - 8.10)}$$

Balance equations for attached phase product mass

net rate of product accumulation	=	-	net rate of product detachment	+	rate of product formation

$$\frac{dC_{Pf}}{dt} = - C_{Pf} \cdot r_{Pd} + C_{Mf} \cdot r_{Pf} \qquad \text{(E - 8.11)}$$

where:

r_{Pf} = specific rate of product formation in biofilm ($M_P \, M_M^{-1} \, t^{-1}$)

Note that we have made or can make the following simplifying assumptions:

- cell or product carbon is not present in the influent;
- the interest of the biotechnologist or the microbial process engineer is generally limited to the steady state process, consequently, the differential terms are often assumed to be zero;
- the relationships according to Monod, between the specific cellular growth rate, μ and the substrate concentration C_S, as generally applied to suspended growth system, is equally applicable to attached growth systems. This has been shown by Trulear (1983 Cellular reproductions and extracellular formation in the development of biofilms, PhD Thesis, Montana State University, Bozeman, MT, USA);

  Thus:

  $$\mu = \frac{\mu_{max} \, C_S}{C_S + K_S} \quad \text{(see Equation 4.28)}$$

- the relationship between specific product formation rate r_P and growth rate as generally applied to suspended growth systems, is equally applicable to attached growth systems.

  Thus $r_P = k_{Pg}\mu + kPn$ (see Equation 4.29)

  where:

  k_{Pg} = growth associated production formation coefficient ($M_P \, M_M^{-1}$)

  k_{Pn} = non growth associated product formation coefficient ($M_P \, M_M^{-1} \, t^{-1}$)

Let us now do some calculations using these relationships.

SAQ 8.4

1) If the biomass concentration in the biofilm is 500 kg m^{-3} and the specific growth rate in the film is 0.5 h^{-1}, what is the specific cell detachment rate if the system is in a steady state? Could you calculate this if the concentration of the biomass in the biofilm was not known?

2) If a system in steady state has a concentration of product in the liquid phase of 400 g C m^{-3} and mass of product per unit area of substratum is 800 g C m^{-2} calculate the specific rate of product leaving the biofilm phase. The following values apply:

   concentration of product in the influent = 0 g C m^{-3}

   dilution rate = 0.2 h^{-1}

   area of biofilm = 1 m^2

   volume of liquid = 1 m^3

   concentration of biomass in the liquid phase = 0 g C m^{-3}

3) The rate of consumption of substrate in the liquid phase has been determined to be 1000 g C m^{-3} h^{-1} when a biofilm:liquid system is in a steady state. If the dilution rate is 1 h^{-1}, the input substrate concentration is 4 kg C m^{-3} and the rate of uptake of substrate by the biofilm is 2000 g C m^{-2} h^{-1}, what is the steady state substrate concentration in the liquid?

8.7.6 Time progression of biofilm parameters

Progression of biofilm parameters shows a great deal of similarity with progression of CFSTR or batch reactor parameters. Progression of biofilm parameters can generally be described by the sigmoidal (S-shaped) curve. The initial lag phase, the slope of the log phase and the level of the steady state are functions of the microbial species used and of the conditions under which biofilm accumulation took place, ie nutrient composition, carbon source, temperature, pH, reactor configuration, hydraulic conditions. Figure 8.8 shows time progression of thickness of biofilms accumulated under ideal laboratory conditions.

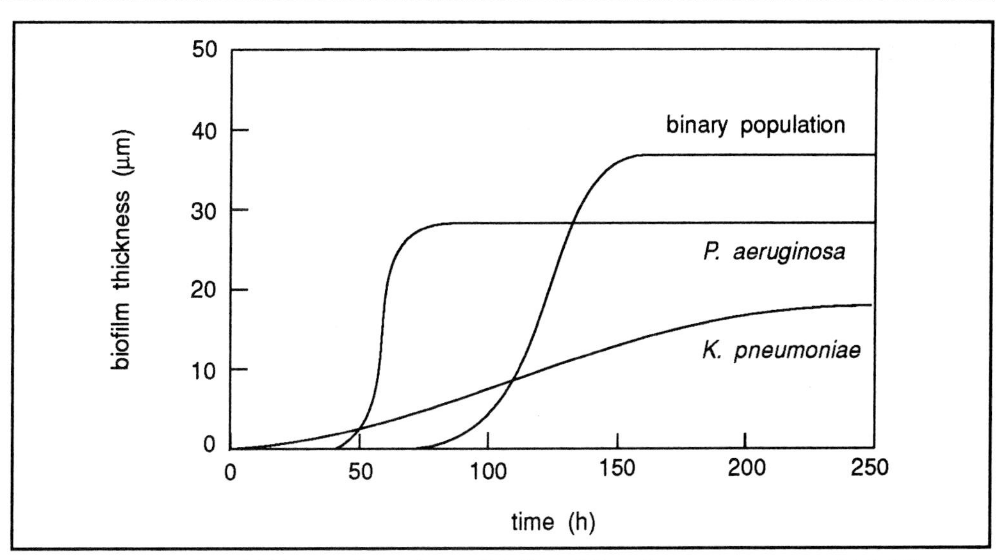

Figure 8.8 Progression of thickness of experimental biofilms of *Pseudomonas aeruginosa, Klebsiella pneumoniae* and a biofilm consisting of both species together.

8.7.7 Specific rates

Information on biofilm specific rates may give an indication of biofilm activity and, as such, is a useful tool in the evaluation of biofilm performance and, hence, plant performance. Specific rates are rates per unit of material involved in the generation of that rate. Various specific rates can be determined in a biofilm, related to production/consumption and/or detachment of cell mass, product mass or substrate. As an illustration, the biofilm specific substrate uptake (consumption, removal) rate is discussed below.

The biofilm specific substrate uptake (consumption, removal) rate (r_{Sf}) is the rate at which substrate is being removed by the biofilm per unit mass of biofilm. r_{Sf} can be evaluated from the material balance for liquid phase substrate (Equation 8.6).

$$r_{Sf} = \frac{D\,(C_{Si} - C_{Sl}) - \dfrac{dC_{Sl}}{dt} - C_{Ml} \cdot \left[\dfrac{\mu_l}{Y_{MSo}} + \dfrac{r_{Pl}}{Y_{PS}}\right]}{C_{Mf} \cdot \dfrac{A}{V}} \qquad \text{(E - 8.12)}$$

If we assume that the concentration of microbial cells in the liquid phase negligible ($C_{Ml} = 0$) and steady state conditions ($dC_{Sl}/dt = 0$), the biofilm specific substrate uptake rate is, therefore, inversely proportional to the mass of biofilm present.

$$r_{Sf} = \frac{D\,(C_{Si} - C_{Sl})}{C_{Mf} \cdot \dfrac{A}{V}} \qquad \text{(E - 8.13)}$$

Equation 8.9 allows further differentiation of the biofilm specific substrate consumption rate function. Assuming no accumulation of substrate occurs in the biofilm, Equation 8.9 can, in steady state form, be written as:

$$r_{Sf} = \frac{\mu_f}{Y_{MSo}} + \frac{r_{Pf}}{Y_{PS}} \qquad \text{(E - 8.14)}$$

Substitution of the relationship between product formation and specific microbial growth rate, a relationship which is applicable both in suspended ($r_P = k_{Pg}\,\mu + k_{Pn}$) and in attached ($r_{Pf} = k_{Pg}\,\mu_{f} + k_{Pn}$) growth, and some rearranging results in:

$$r_{Sf} = \mu_f \cdot \underbrace{\left[\frac{1}{Y_{MSo}} + \frac{k_{Pg}}{Y_{PS}}\right]}_{a)} + \underbrace{\left[\frac{k_{Pn}}{Y_{PS}}\right]}_{b)} \qquad \text{(E - 8.15)}$$

According to Equation 8.15 substrate consumption by the biofilm has two components: one component, a) is growth rate dependent and concerns the conversion of substrate into cell mass (Y_{MSo}) and product (Y_{PS}), the other component, b) is growth rate independent and concerns only product formation.

SAQ 8.5

1) If the specific rate of substrate consumption by a biofilm is 3.834 g substrate C/g biomass h and the observed growth yield is 0.3 g biomass C/g substrate C, what is the product yield given that the specific rate of product formation is 0.2 g product C/g biomass C h^{-1}? The specific growth rate of the organism is 0.5 h^{-1} and all of the product formed is growth linked.

2) Combine the observed growth yield value given above with your calculated product yield. From this combination, can you conclude that substrate is only converted to biomass and product?

8.8 In retrospect

In Chapters 4-8 we discussed in detail the main reactor types encountered in bioprocess engineering Their main characteristics were described and for some we addressed their practical application. Most attention, however, focused on the mathematical description of the reactors resulting in models describing concentration changes as a function of time under various process conditions.

It cannot be over-emphasised that these models were derived at under ideal conditions, meaning that all circumstances that were critical in arriving at the models were known exactly and could be precisely controlled. Only then was it possible to generate smooth and predictable curves.

It must be clear to the reader that ideal conditions never exist in practice, not even under (relatively) well controlled laboratory conditions, let alone when operating full scale plants. The main objective of discussing idealised reactor configurations was to give the reader insight into functional and operational aspects because only when the principles are fully understood can one attempt to fruitfully approach the practical side of reactor configurations used in bioprocess engineering.

Summary and objectives

In this chapter we have examined attached growth reactors. First we described what was meant by attached growth reactors and explained the advantages, particularly in relation to the dilution rates that could be achieved, using attached growth systems. We also briefly made a comparison between using immobilised cells and enzymes. Physical and chemical methods for immobilising cells were outlined together with the ways in which immobilised systems are used in practice. The major part of the chapter was devoted to examining the balance equations for attached growth systems. This part of the text was based upon a description of the processes involved in spontaneous cell immobilisation. The model used was to distinguish parts of the immobilised biofilm reactor: the biofilm and the liquid. Questions, designed to test understanding of the derived equations were also included.

Now that you have completed this chapter you should be able to:

- use a wide variety of terms employed in the description of the processes taking place in biofilm reactors;

- explain the advantages (and disadvantages) associated with using attached growth systems;

- list the main methods by which cells may be artificially immobilised;

- identify the factors which influence the formation of gradients in biofilms;

- use balance equations relating to biofilms to calculate a wide variety of parameters from supplied data.

An introduction to process control

An introduction to process control

9.1 Introduction

In the previous chapters, we have examined the performance of reactors associated with biotechnological processes. In the remaining part of this text we will examine process control. To do this we have divided the text into several short sections to enable you to pace your rate of study to suit your needs. In the first part we will examine basic control schemes. Subsequently we will discuss a variety of controllers and non-linear control before moving on to exploring instrumentation and aspects of optimisation. These sections will provide you with an overview of the potential of modern process control and knowledge of the most important tools used by control engineers. Our primary objective is to enable you to develop an awareness of the potential and difficulties of automated process control. We begin by examining what process control is and how process control was developed.

9.1.1 Process control

early approaches to process control

Process control forms a part of process optimisation. In the early days, there was no need for close process control (due to the moderate competition). Nor was it really possible because of the lack of appropriate instruments. Also very little theoretical knowledge was available. However, it was soon discovered that besides the efficiency of the selected strain, the initial conditions had a profound influence on the product quality and process efficiency. Therefore, initial conditions were optimised by trial and error. Of course, this way of process optimisation took a long time. The early concept of describing microbial growth in mathematical terms (the early 1950s), made people realise that optimal initial conditions could be calculated (at least a good starting point could be produced). Laboratory experiments, transformed into a process model, formed the basis for these calculations. In this way, optimal initial values or working points could be obtained much faster, and the development time of new processes was reduced. Therefore, the use of models forms a most attractive starting point for the optimisation of a process.

Besides the choice of the initial conditions, it is recognised today that close process control can further improve the process economics. This especially holds for processes with a high degree of complexity.

A clear advantage of an efficient control strategy is the more constant product quality. This will yield either a higher price for the product or lower downstream processing costs.

Quality Assurance

In the case of many products, but especially medicines, authorisation to sell or use the product is only granted if appropriate Quality Assurance is maintained (Quality Assurance and market authorisation of medicines are dealt with in detail in the BIOTOL text 'Biotechnological Innovations in Health Care'). Such Quality Assurance issues demand an efficient control strategy for the processes involved in production.

There is a growing tendency for process control to be integrated. For example, nowadays, the control of the fermentation step is often combined with the control of the downstream processing steps. This may yield further economisation of the complete production plant. Finally, due to efficient process control, processing problems can often be detected in an earlier stage and their origin may be located easier. Again, this is of special interest as the complexity of the processes increases.

optimisation

At this stage we should distinguish between large scale bulk production (yeast, ethanol etc) and small scale production (chemicals, monoclonal antibodies, restriction enzymes, etc). In large scale processes substantial investments can be made in order to optimise each specific production process. Considerable attention should be paid to processing strategy and efficiency because of the relatively low added value of the products compared to the price of the raw materials. Even if the production costs can only be reduced by a small percentage, it will still be worthwhile to optimise the process further by means of on-line process control. This is of course due to the scale of the process. For each process, specific control strategies can therefore be developed.

flexibility

Small scale production does not allow such an approach. Normally, the bio-catalyst and the media used are less than optimal. Therefore, most investments in process optimisation will be made on this part. Nevertheless, on-line control can still be beneficial, and is sometimes even crucial, because a process may be unproductive or even terminate due to unfavourable conditions. Furthermore, media may be very expensive (likely for animal cell cultures); a sound control strategy may reduce medium usage. Therefore, for small scale production, control strategies must be developed that are widely applicable, without the necessity of being completely optimal. Flexibility is the key word here.

It may be worthwhile to combine control of the primary fermentation with the control of other parts of the production plant. For some newly developed processes this may even be essential, as part of the downstream processing forms an integral part of the fermentation step. In these situations, fermentation control and control of downstream processing will always need to be integrated.

Thus the key elements in process control are different depending upon the scale of operation. With this introduction in mind we will turn our attention to the range of basic control schemes that are available to us.

9.2 Basic control schemes

A number of basic control schemes can be distinguished. Each basic scheme serves a special function. For complex control problems, we will often have to combine these basic control schemes in order to obtain acceptable controller characteristics.

importance of an input

In order to control a process, we must be able to influence the process by means of its input, ie variable(s) which can be used to influence the process. For example, we might be able to influence an aerobic process by altering the oxygen tension.

This may seem very trivial, but some processes cannot be controlled due to a lack of input, eg an anaerobic batch fermentation cannot be controlled except by changing pH and temperature.

We can identify four basic control schemes. These are:

- feedback;
- feedforward;
- cascade;
- inferential.

We will examine each in turn.

9.2.1 Feedback control

output

setpoint

When we speak about process control, we normally mean feedback control. In this type of control, measurements from the process are used to calculate a suitable control signal: the (on-line) measured values are fed back to the process input. The scheme of feedback control is given in Figure 9.1. If we take a closer look at this figure, we see that the measured value (the process output, the actual process value) is compared to a desired value (the setpoint value). The difference between these two values is fed into the controller block, which uses this difference to effectively adapt the input signal. Therefore, we are able to suppress the influence of (unexpected) system behaviour or disturbances by means of feedback control.

A feedback controller responds to disturbances and therefore reduces deviations from setpoint.

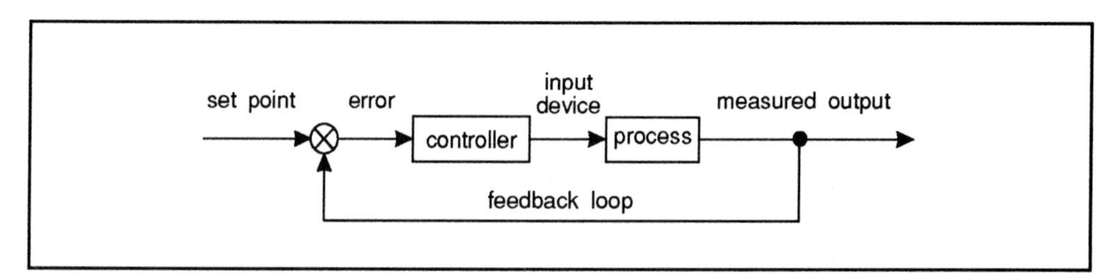

Figure 9.1 Feedback control.

The terms 'input' and 'output' are sometimes confused by students. Before we continue with our explanation of feedback control let us consider these two terms a little more fully.

Input variables are independent variables, while output variables are the dependent variables. Of course, what exactly is considered an input (or output) variable depends on the view that a person has on the system. For example, the temperature will be an output variable with respect to temperature control, while the flow rate of cooling water may be the respective input. The same temperature (or its setpoint value) can, however, be regarded as an input signal for the process (when the temperature is actually used to modify process characteristics) or, more often, it can be regarded as a parameter (fermentation at constant temperature).

Finally, what is input for one part of the process may be output for another part (for example, the process 'output' serves as 'input' for the controller).

Now let us return to feedback control.

Due to the feedback loop, as illustrated in Figure 9.1, process stability may be improved by feedback control, because the controller can respond to an increasing signal by suitable action. Let us illustrate this with the following example.

Let us assume that the 'process' illustrated in Figure 9.1 is a culture of micro-organisms growing on glucose. Let us also assume that for the best process economics we should maintain the concentration of glucose in the culture medium at between 5 and 6 m mol l^{-1}. The input device in this case is a pump which pumps glucose solution into the process (ie the input is a solution of glucose). The glucose in the output from the process is measured (eg by a glucose electrode) and its signal is sent back to the controller. The set-point value is arranged such that if the glucose level in the output falls below 5 m mol^{-1}, then the controller switches the pump on and glucose solution is pumped into the process. If the glucose level rises above 6 m mol^{-1} then the pump is switched off. Thus, in principle the level of glucose is maintained between 5-6 m mol l^{-1}. This type of control is relatively straightforward. In practice, however, there are many pitfalls.

Π See if you can write a list of possible problems with feedback control. To help you think about this, consider what would happen to the glucose concentration in the outflow if there was a very long time delay between the measurement of glucose in the outflow and the setpoint device switching on the glucose pump. Or, what would happen if the controller had a very large pump or the glucose in the input glucose solution was very concentrated?

We anticipate that you would predict that if there was a long time delay in the pump being switched on, then the glucose concentration in the output would continue to decline below 5 m mol l^{-1}. If the pump was too large, we would expect that once it was switched on, so much glucose solution would be pumped into the process that the glucose concentration in the process would rise very rapidly above 6 m mol l^{-1} before the detector in the output could respond.

Graphically we could represent this in the following way.

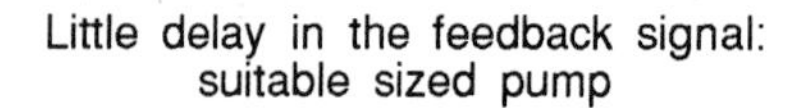

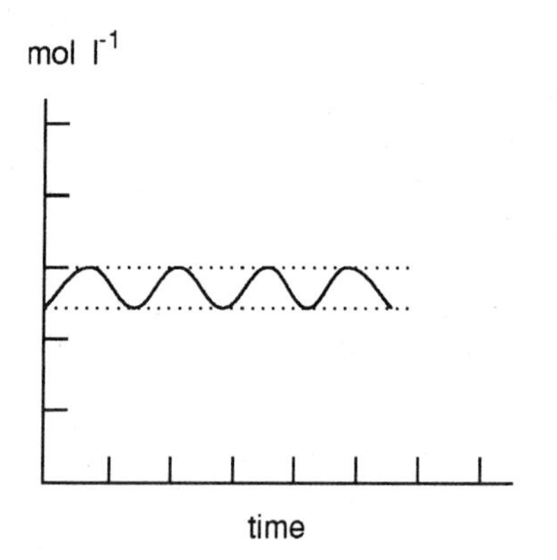

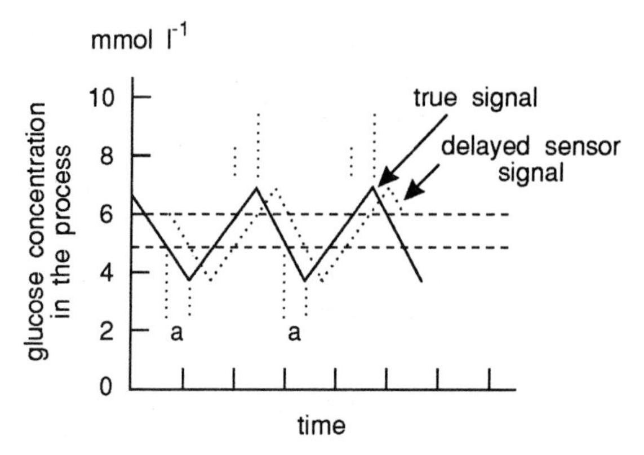

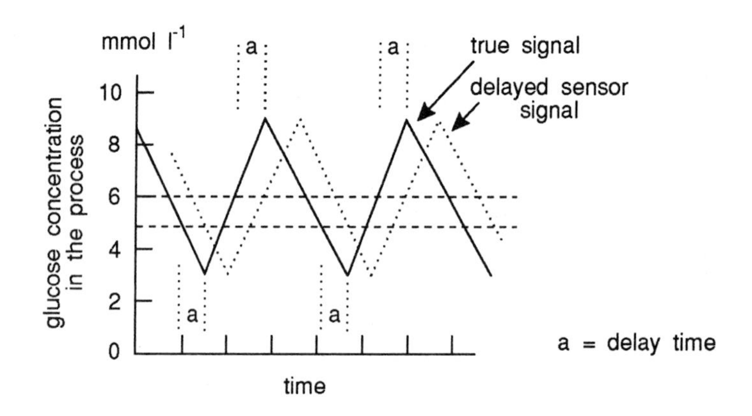

Note how the glucose concentration oscillates in each case.

We can generalise what we have learnt from the above example, in the following way.

overshoot
delay time
dead time

When a signal from the measured output deviates from the set point, this deviation will result in a modification of the input signal by the controller. The effect of this, however, may not be apparent on the system output before some time has elapsed (due to the system dynamics). Therefore, the deviation may still grow for some time, causing the controller to modify the process input even more. As the result becomes apparent it may become clear that the total modifications on the input were too strong. A deviation to the other side may be the result. This common phenomena is termed 'overshoot'. It is often associated with systems where the input signal does not immediately effects the system output. Therefore, it commonly occurs in systems with a delay time (also termed dead time): a period in which the system does not respond to a change in the input signal. Note overshoot can occur for other reasons. For example in slow response systems. In slow response systems there may be very little delay in the signal being received by the controller, but the response of the controller or input device might be slow to develop.

If the overshoot becomes too large, the controller may destabilise the process. The overshoot then becomes larger than the original deviation. This, in turn, will cause an even stronger controller action, resulting in an increasing overshoot. The process starts oscillating with an increasing amplitude. This may occur if a 'too strong' feedback is used for systems with a long dead time. Therefore, care must be taken, in order to avoid these situations. We will discuss this in more detail in a later section.

Feedback controllers are often used for temperature and pH control. The measured temperature is then compared to its desired value (its setpoint value) and the fermenter is heated or cooled accordingly.

SAQ 9.1

A bioreactor is set up with an oxygen control device. This consists of an oxygen electrode which measures the oxygen in an outflow from the reactor and an air pump with a setpoint controller. The setpoint controller is set to switch the pump off is the oxygen tension in the outflow is above 8% saturation and to switch the pump on if it falls below 5% saturation. A graph of the actual oxygen tension in the outflow is provided below.

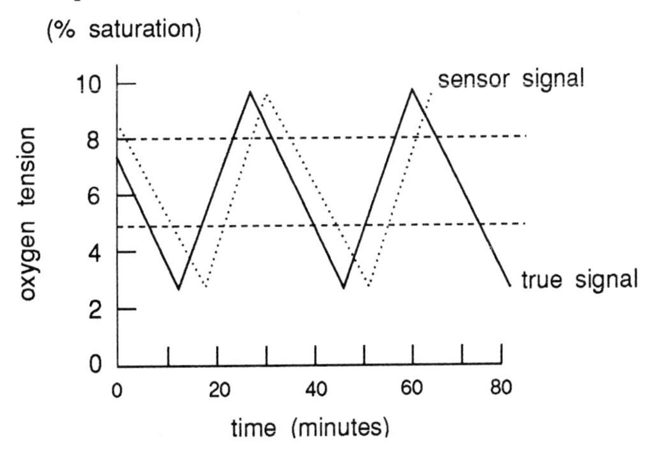

From this data:

1) Calculate the likely delay time in the oxygen control loop.

2) Explain what might happen if the air input was replaced by pure oxygen.

3) Explain how the operators might achieve an oxygen tension to always fall between 5-8% saturation.

9.2.2 Feedforward control

Let us assume that the oxygen concentration of a batch reactor is to be kept at a certain minimum value in order to prevent oxygen depletion and, at the same time, to keep the operational costs as low as possible. As we have just learned about the principle of feedback control, the measured oxygen concentration is compared to its setpoint value and the difference is used to adapt the oxygen supply rate (eg by means of controlling an air pump; adaptation of the stirrer speed will support the control process.

At the start of the experiment a high oxygen concentration is measured (the small amount of bacteria consume little oxygen). Therefore, the air pump is set to a minimum value. Due to exponential bacterial growth, the oxygen demand will increase, and therefore, the oxygen concentration drops. At the moment that the oxygen concentration reaches its setpoint value, the controller starts functioning. From this point on, the controller will have to increase the oxygen supply rate exponentially in order to keep up with the exponentially growing oxygen demand. Here, we see clearly a disadvantage of the feedback controller: it can only respond to observed deviations. In order to increase the oxygen supply rate exponentially, the deviations must increase as well. Of course this is not very satisfactory. The controller will always react too late.

Clearly, we need another approach. First of all, we know that the oxygen demand will increase exponentially. We can therefore anticipate and increase the air flow rate accordingly (on the basis of some growth relations). If no disturbances occur (our growth relations were correct etc), this scheme will yield precisely what we want: a constant oxygen concentration at the specified level. This way of anticipating is called feedforward control. The set up is given in Figure 9.2. Feedforward control can be applied without measuring the variable of interest (the oxygen concentration). The same controller set up can be used in controlling the substrate concentration of a fed-batch fermentation process.

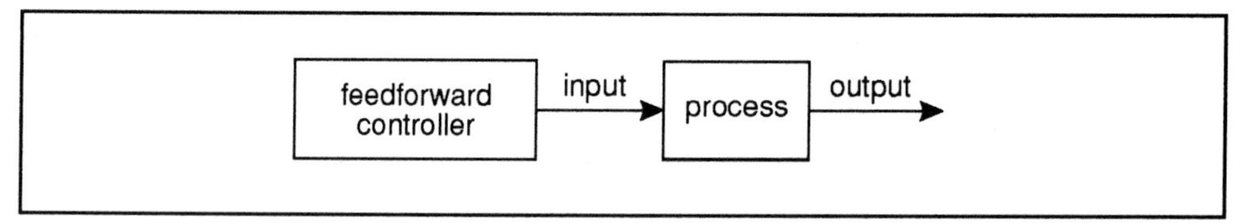

Figure 9.2 Model based feedforward control.

Another type of feedforward control is illustrated in Figure 9.3. Here, the controller acts on the basis of measured disturbances. Let us assume that the temperature of a reactor is to be kept at 30 degrees. Normally, the cooling water has a temperature of, say 15 degrees. However, a shift in the temperature of the cooling water to 20 degrees is observed. We can then act immediately and modify the flow of the cooling water accordingly.

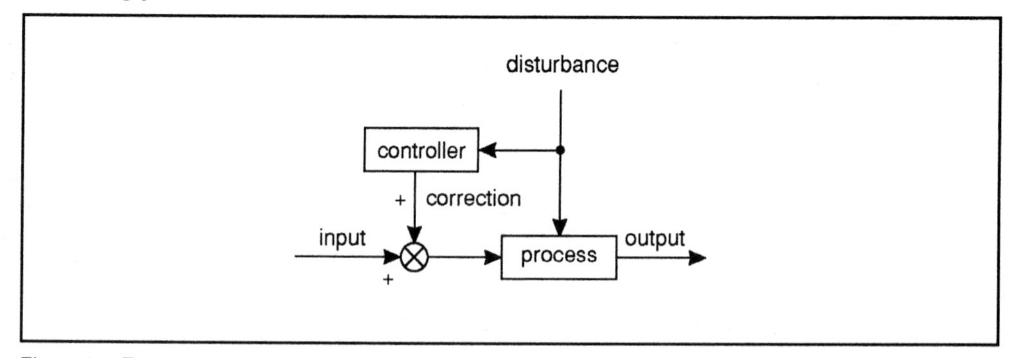

Figure 9.3 Feedforward control with disturbances.

Usually this type of feedforward control can be set up in advance. Thus we can build into the control a relationship between the extent of the disturbances and the correction factor. It is important to realise that feed forward control is model-based.

Often it is wise to combine feedforward control with feedback control in order to be able to compensate for small model deviations and unpredicted (unknown) disturbances. This is schematically represented by Figure 9.4.

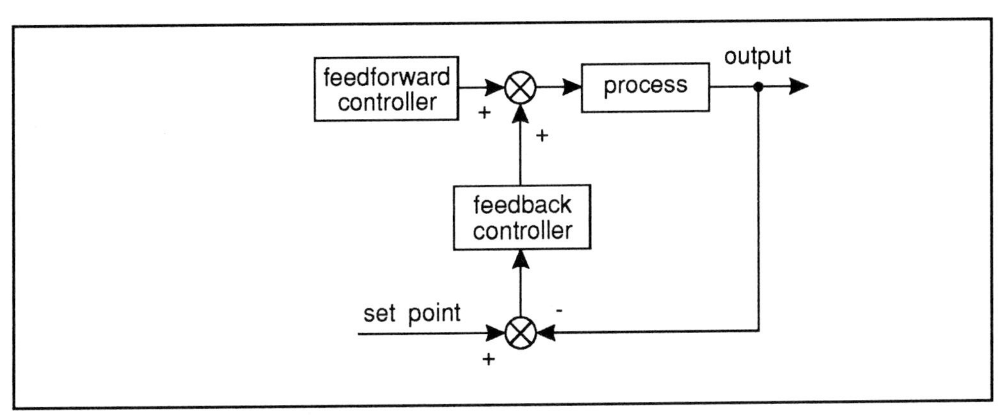

Figure 9.4 Combined feedforward and feedback control.

Thus to our oxygen controller described above and illustrated in Figure 9.2 we would also add a feedback loop, which would result in the situation depicted in Figure 9.4. The feedforward controller would enable us to establish general oxygen tension in the process according to a model. The feedback controller would allow us to refine the actual level of oxygen in the process if it deviated from the model (calculated) value.

9.2.3 Cascade or supervisory control

The term cascade control indicates a special set up of different control loops: an outer control loop dictates the setpoint of an inner loop. This set up is used to make a controller more suited for start-up periods. The use of feedforward control reduces the need for cascade control. However, cascade control can be implemented simply, by means of some standard analog devices, while feedforward control usually involves a digital computer and a model of the system.

primary and secondary controller

Cascade control is very efficient if the process can be split up into a fast and a slow part. The set up that is applied then is illustrated in Figure 9.5, in which the inner loop (with the secondary controller) has the smallest time constants (the fastest process part). The outer loop generates a setpoint for the inner loop.

9.2.4 Inferential control

Although the feedback and feedforward schemes offer some control, we may still not be able to control many biotechnological processes well. This is mainly due to the fact that the most important process variables (eg the biomass, substrate and product concentration) cannot be measured on-line. Direct feedback control in which the variable of interest is measured directly is then not possible.

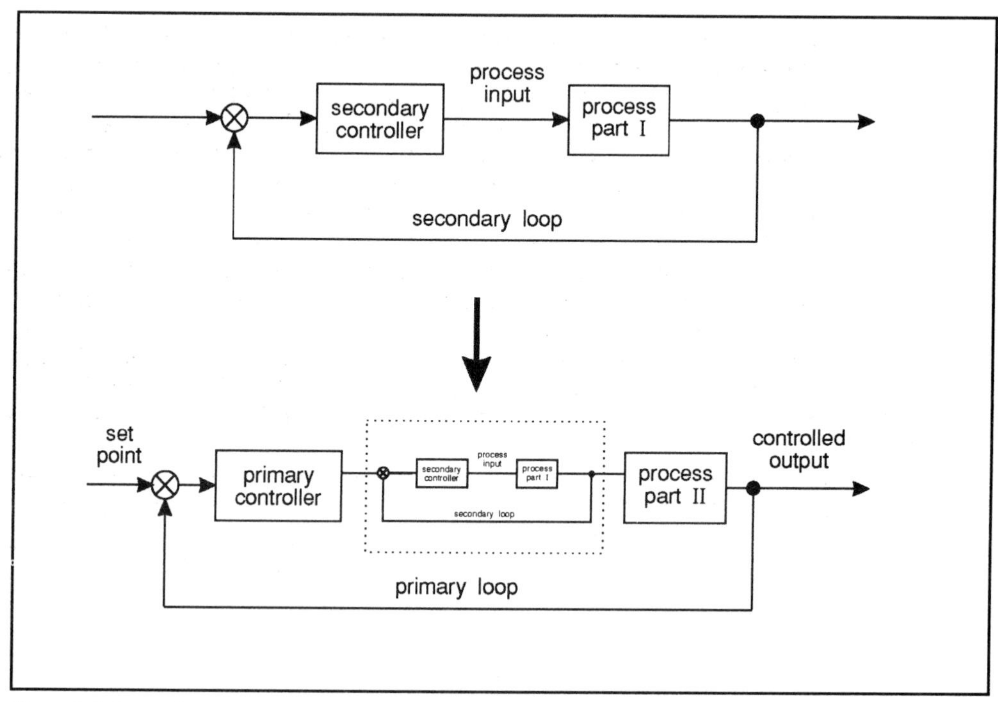

Figure 9.5 Cascade control.

There are two ways to overcome this problem. One is to control a secondary variable instead of the primary variable of interest. We might regard these as indirect (or inferential) methods. For example, we might use oxygen uptake rate as an indirect (inferential) method of monitoring substrate consumption rate.

RQ control

Frequently the state of a culture is measured by the respiratory quotient (RQ) of the culture. The respiratory quotient is determined by the ratio of CO_2 generated to oxygen consumed. The RQ can be derived on-line by calculating it from CO_2 and O_2 measurements in the exhaust gases. Clearly the rates of CO_2 production and the rate of O_2 utilisation depends not only on the rate of metabolism but also on the nature of the substrate used to support this metabolism. Since RQ can be determined directly on-line, RQ control can be operated. Note the reason why such methods are referred to as inferential methods. In the case just cited, we 'infer' the state of a culture from the RQ value.

Although such methods have found wide application they do have their drawbacks. One such draw back is that we assumed that the value we are using (eg RQ or turbidity) is directly related to the parameter we wish to control in a predictable way. This might not always be the case. For example, we might use turbidity to control biomass concentration on the assumption that the turbidity of a culture is directly related to the biomass concentration. What happens if the cells change shape? For example they might produce spores or become filamentous. Is the turbidity related to biomass in exactly the same way for vegetative cells, filaments and spores? The answer is no.

There are, however, ways round such problems. Let us take an example in which we wish to control the substrate concentration in the output.

state estimators

If the substrate concentration needs to be known continuously, the control scheme could be improved by using a state estimator. State estimators (or observers or filters as they are sometimes described) use a model to calculate variables based on the on-line measurement of one (or more) other variables. Thus modelling as well as one (or more) on-line measurements form the basis of the algorithm that estimates the state of the process. State estimators are discussed in more detail in Section 9.5. A feedback controller that uses the results of such a state estimator and compares the estimates with its setpoint is also called an inferential controller. Its scheme is given in Figure 9.6. The state estimator interacts with the controller and the measurement device. It is not considered an ordinary feedback controller, because special care must be taken in order to ascertain that the estimated values are a good representation of reality. This is not only a problem of the estimator, but also of the controller, and the process itself. All elements in the loop interact with each other. Although powerful, inferential controllers are quite complex, they must be tuned carefully and extensive testing is usually necessary.

careful tuning and testing required

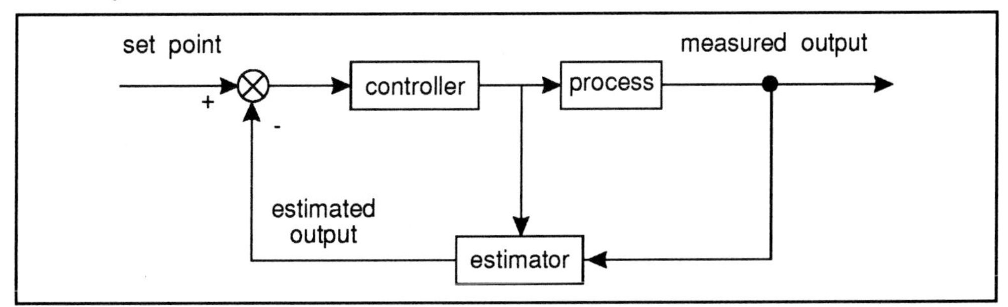

Figure 9.6 Inferential control.

9.2.5 System identification

As we move on from simple controllers to the more complex ones, it will become clear that the use of a process model becomes increasingly important. This especially holds for inferential control, as its functioning is completely dependent on the functioning of the state estimator, which in turn, is dependent on the quality of the provided model. Models are never exact however, and this is particularly important in biotechnological processes. In order to keep up with (non-modelled) changes in the process, models should be updated regularly (at least the model correctness must be tested regularly). This can be accomplished by any kind of on-line state estimator and a system identification unit. The general control scheme with a system identification block is described in Figure 9.7.

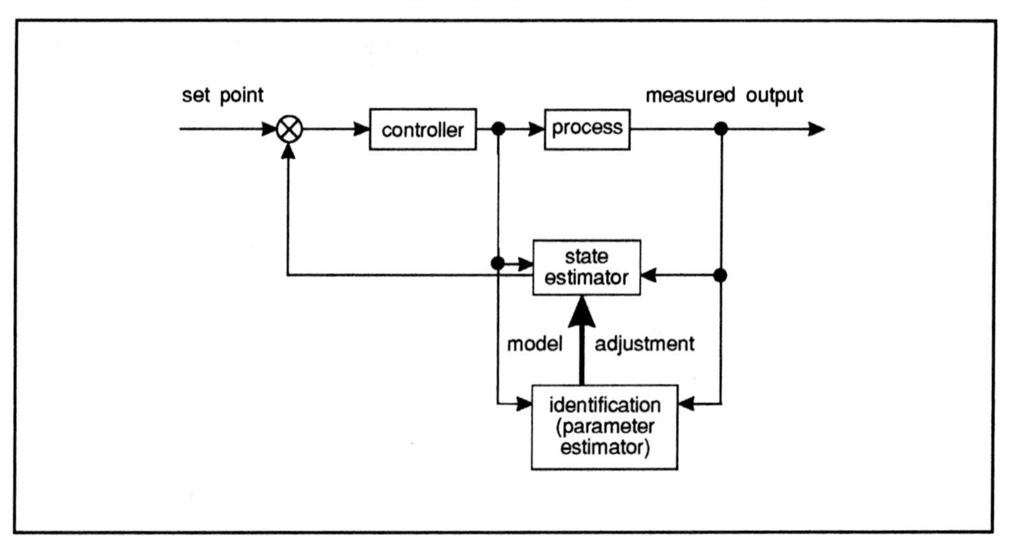

Figure 9.7 Process control with system identification.

adaptive controllers

A controller that involves any kind of on-line system identification is called adaptive, as it adapts the underlying model to new situations.

Usually, the development of a successful adaptive structure involves extensive trial and error (by means of simulations and experimental tests). Sometimes, however, it is possible to prove mathematically the convergence of the parameter estimation procedure. Normally, this is a difficult task, which nevertheless must be carried out in order to avoid problems of faulty or divergent parameter estimation procedures.

9.3 Systems dynamics

In this section, we will briefly discuss the subject of system dynamics. Because the dynamical behaviour of a process is of importance in process control. Special process dynamics demand specially tuned controllers. If a controller is not well designed for a given process, the controller will function badly, resulting in a high overshoot, a large settling time or even instability of the controlled variable.

oscillation in the measured output and input

Let us examine this a little more closely. Look back to our discussions of feedback control in Section 9.2.1. You will recall that we discussed how the concentration of glucose in a process is dependent upon the delay time in the feedback loop, the size of the pump and upon the glucose concentration in the input to the process. We discussed this purely in a descriptive manner. Systems dynamics is a much more rigorous approach to describing such features of a system. Clearly such an approach is important in the design and development of a system. Systems dynamics for example enable us to calculate (predict) oscillations in process variables and whether to estimate a process will be stable. Another way of considering systems dynamics is to recognise it as a method of analysing the effects of changes in one variable of a system will have on other variables. If you think about a system, oscillations in a measured variable in the output of a process will, in a feedback loop, cause oscillation in the input. We can represent this in the following way:

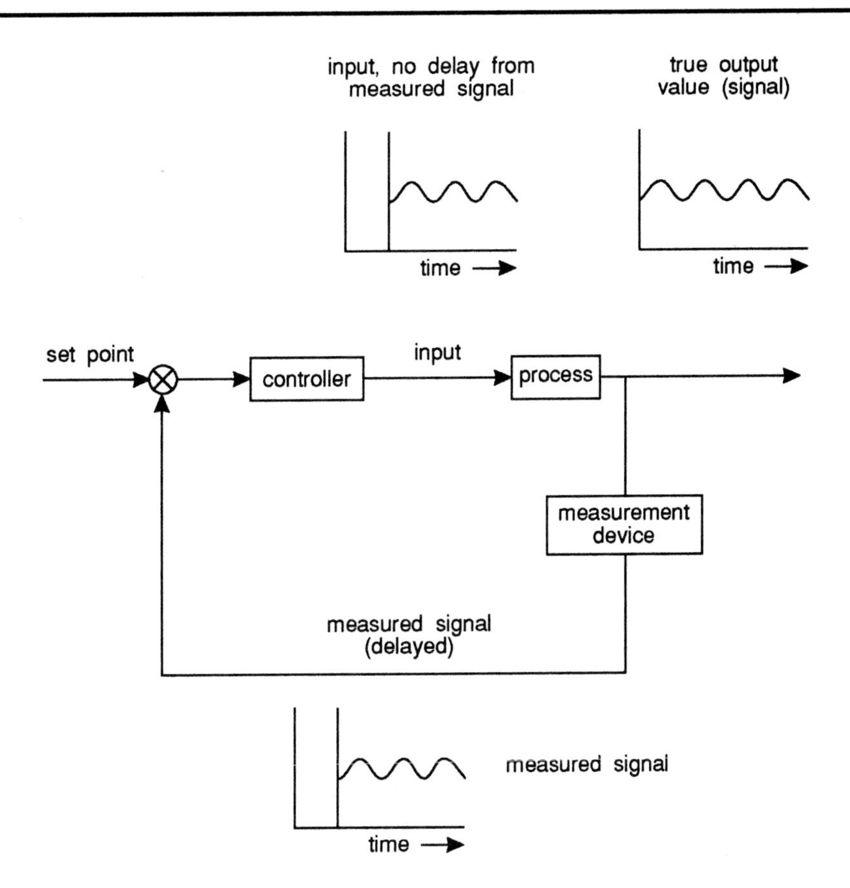

We can ascribe to these oscillations, two basic properties:

- their frequency;
- their amplitude;
- phase shift ('delay').

The frequency and amplitude are of course features of the system as a whole. Most commonly the delay is a feature of the process or the measuring instrument. In most instances signal transfer from measuring device to controller is fast and occurs with little delay.

The systems dynamics approach enables us to make calculations to establish the amplitude and frequencies in change in the value of variables within a system. Mathematically these are not, however, easy to handle requiring s-domain/Laplace transformation.

Systems dynamics demand quite sophisticated mathematical knowledge and skills. Such knowledge is not available to all those who need to work in liaison with process technologists. For this reason, we have split the discussion of systems dynamics into two parts. In the core text, we have confined ourselves to a descriptive discussion of the principles involved. For those who need a more rigorous and numerical approach to this topic, we have provided a more detailed mathematical approach within Appendix 2 at the end of the text.

9.3.1 Some key terms in systems dynamics

SISO

In biotechnological processes there are many inputs, for example oxygen supply rates, stirrer speed, flow rate of cooling water, temperature of cooling water etc. In many instances however, a single input and single output are considered. Usually the most logical input is chosen. For example temperature may be controlled by the flow rate of cooling water. Such processes in which only one input and one output is considered are often referred to as SISO (single input single output) systems. Examples may vary from the temperature control of a fermenter to the control of the substrate concentration in a fed batch fermentation. The measured variable (eg temperature, CO_2 production rate) is then controlled at the setpoint by using one variable that is regarded as the single input variable (eg flow rate of the cooling water, substrate addition rate).

The case of glucose input into a culture of micro-organisms described in Section 9.2.1 is an example of an SISO system.

However, we are often dealing with more complex processes. In these there may be more than one input or output from a process and these may need to be controlled at the same time.

MIMO

If the different control loops cannot be studied separately, or if the number of inputs does not equal the number of process outputs, we are dealing with MIMO (multiple input multiple output) processes.

time dependent and time independent systems

Another important property is whether or not the process parameters are varying with time. We can thus describe systems as being time independent systems or time dependent systems.

Time dependency may cause some problems in process control. Especially, the development of analytical solutions for time dependent systems is frustrating, as the exact form of the time dependency is often not known.

9.3.2 The importance of linear systems in systems dynamics

In practice, it is not always necessary to investigate the system behaviour as a result of all possible input signals. This especially holds for systems which behave in a linear way (linear systems). For linear systems, the principle of superposition holds. Let us explain what this means. We will use the symbol I(t) to represent an input signal at time t and O(t) to represent an output signal at time t; in a linear system I(t) is proportional to O(t). In other words if input signal $I_1(t)$ results in an output signal $O_1(t)$, and the input signal $I_2(t)$ causes an output signal of $O_2(t)$, then for linear systems, the input signal $a_1.I_1(t) + a_2.I_2(t)$ will cause an output signal of $a_1.O_1(t) + a_2.O_2(t)$ (a_1 and a_2 are real constants and I_1 is one input and I_2 is a second input etc). The effect of different input signals may be added. Therefore, linear systems can be easily described and analysed.

linearisation point

Non-linear systems are, as the term explains, systems for which the superposition principle does not hold (exactly). In fact all real processes are to some extend non-linear. However, many non-linear systems can effectively be linearised. The linearised process equations are then (more or less) good approximations of the non-linear equations. This approximation only holds in the neighbourhood of the working point from which the equations were linearised; the linearisation point. Results obtained from a linearised system description are only locally valid, and only for input signals that are not too large.

The superposition property allows one to calculate the dynamic behaviour as a result of all input signals from just one probe signal (often a step or an impulse signal). Because linear systems can be more readily described and studied, and because most non-linear systems can be linearised effectively, most of the control theory concerns linear systems.

A mathematical approach to systems linearisation is given in the Appendix 2.

9.4 Basic controller actions

The most comprehensive controller is an on/off controller. Unless the switching frequency is high relative to the characteristic times of the process, and the magnitude of the hysteresis can be kept small, this type of control will not be satisfactory. Nevertheless, it is often used, due to its mechanical simplicity, when the demands on the controlled signal are not too strict (eg the temperature control of a water bath). We have introduced the term hysteresis. Hysteresis in this context means the extent to which the controller does not alter the process input signal. This of course will be reflected by differences in the actual value and the set value.

Π Make a list of as many on/off controllers used in the home as you can think of.

After a little thought you will probably have been able to make quite a long list. Here we cite some typical examples:

thermostats: used to maintain a particular room temperature, or the temperature in an oven, or the water in a washing machine etc all tend to be on/off switches. Similarly, the thermostat coupled to gas powered boilers are also usually on/off systems.

alarms: such as in an alarm clock, the alarm is either ringing or not ringing and is therefore controlled by an on/off device.

timers: on ovens, central heating devices, video recorders etc. These too are usually on/off controllers.

You should have concluded that on/off controllers are very common. We did, however, imply in the beginning of this section that on/off controllers are only satisfactory if the frequency they can switch on or off is high relative to the typical time of the process.

Π Draw a graph of what would happen to the temperature of a room if the thermostat controller used in a room could only be switched on or off maximally at a rate of once per hour and there was 1 hour delay between the controller receiving the signal and switching on the heater.

You should have realised that the temperature of the room would fluctuate. Effectively the slowness of the thermostat to switch represents a delay in the feedback of the temperature signal to the heater. Your graph should thus have looked like the one drawn below. Note the overshoot as the thermostat controller fails to switch the heater off or on until 1 hour has elapsed.

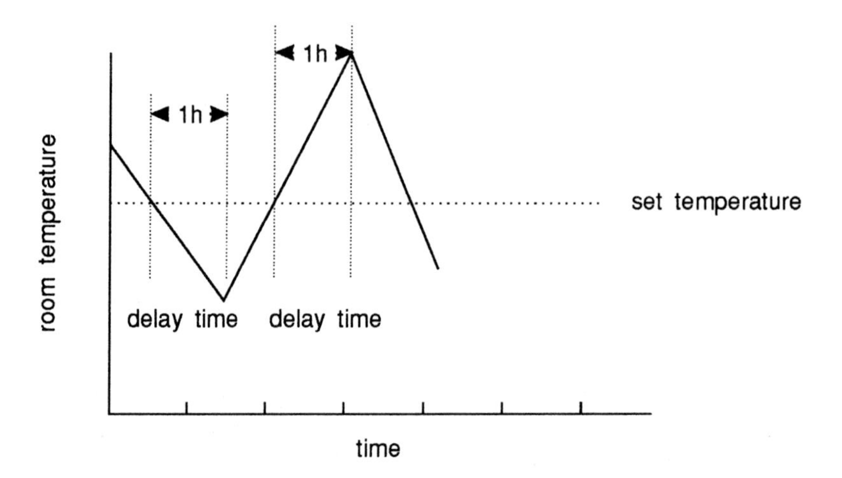

We have produced a graph of temperature against time of a room using a thermostat with a built-in delay of 1 hour.

Π What would happen if the thermostat could only be switched on or off once per day (ie a very, very slow switching frequency)?

The temperature of the room would, of course, only be thermostatically controlled for a very small part of the day. We would then have to wait a further 24 hours before the thermostat would respond again. This is a rather far-fetched example but it does show that for on/off controllers, to keep constant (controlled) values of parameters, such controllers are only effective if they have short time delays and a high frequency of switching.

Π What would happen to the temperature in the room if the slow thermostat controller switched on a very powerful heat?

You should conclude that the temperature of the room would greatly overshoot (see Section 9.2.1) and it might take a very long time after the heater has been switched off for the room to cool back to the set temperature. Graphically we can represent this as:

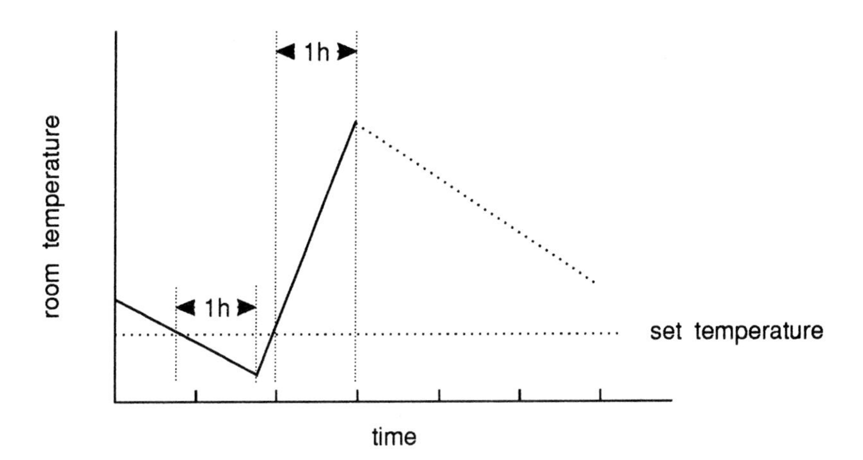

We can conclude therefore that on/off controllers, although simple to design and operate, are only appropriate if their switching frequency is relatively high and the device they control (magnitude of the input) is appropriate for the system.

This discussion has shown us that in selecting an on/off controller we must very carefully consider the frequency it can be switched and the appropriateness of the response that the on/off switch produces.

In many processes, however, the required response is variable and simple on/off controllers do not give a sufficiently refined response. What may be required in such circumstances are proportional controllers. Proportional controllers are controllers which will generate a continuous range of input values to control a process. In the next section, we will introduce you to the most common proportional controller systems.

9.4.1 Proportional controllers (P)

A proportional element simply multiplies its input with a constant factor (denoted K_p). If a proportional element is used to control a process, the controller output (which is used as the process input) equals the controller input (the difference between setpoint and process output), multiplied by a constant factor K_p.

Thus:

controller output signal = controller input signal x K_p,

or:

process input signal = (process output signal - set point)K_p

since the controller output signal is the process input signal and the controller input signal is the process output signal - set point.

In practice the process input signal is also magnified. We can thus represent the system as show in Figure 9.8.

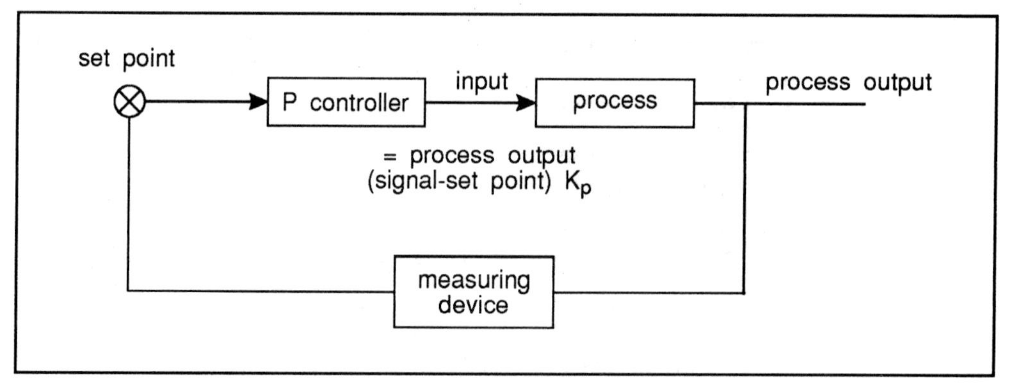

Figure 9.8 Feedback loop with a P controller. Note the process input signal = (process ouput - set point) K_p. See text for description.

In controlling this process, what we seek to do is to keep the process output constant (ie at the set point). If the measuring device detects a deviation of the output from this set point, this generates a signal which is detected by the controller (the controller input signal). The controller generates an output signal which causes the process to respond thus modifying the process output to bring it closer to the set point. Hence in such a system, changes in the processes deviation from the set point generates a change in the controller's input signal which in turn, generates a change in the controllers output signal. For many processes, such systems can, in principle, maintain the process output at the set point value with very little error.

Let us examine such a system a little more closely. In normal process conditions an input value is generated which should give the desired process output (setpoint value). This input can be calculated by means of a process model or obtained experimentally. It matches the process output to the desired setpoint and can be regarded as a constant feedforward part of the controller.

A deviation from setpoint is necessary to obtain a deviation from the constant input signal. Therefore, if the constant input does not exactly match the setpoint value, the controller will not be capable of keeping the output exactly on its setpoint. A steady state error will be the result. The larger the K_p, the smaller this effect will be. On the other hand, a large K_p results in a nervous controller. We mean by nervous that even small deviations from setpoint will result in a relatively large controller action.

Π Make a list of the problems that might arise from a nervous controller.

The sort of problem you might have envisaged are:

- that minor measurement errors will lead to a relatively large controller action, in other words a large K_p value will tend to amplify the errors made in measurement;
- mechanical wear to valves as a result of rapid valve adjustments;
- even small delays cause large overshoots;
- drift in sensors.

drift

Consider a system measuring the flow rate in a pipe and that we wish to keep the flow rate constant. The sort of scheme we have in mind is:

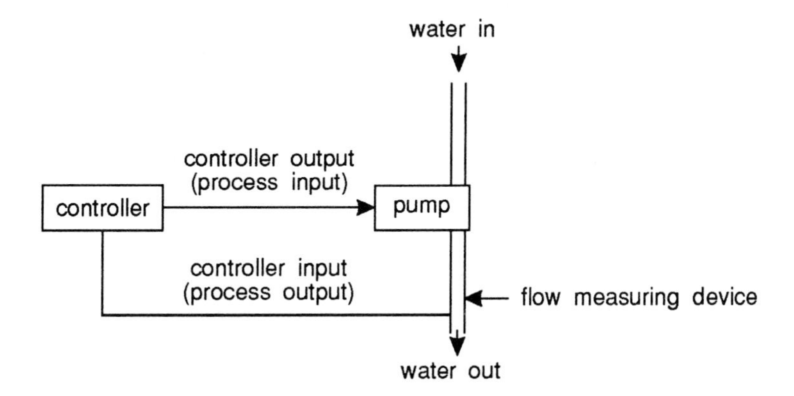

Thus, if the flow measuring device detects that the flow rate is too low (ie below the set point), it causes, through the controller, the pump rate to increase. The pump will however show some wear and its pumping efficiency will slowly change as a result of this wear. We call such a slow but progressive change -drift. A high K_p value will reduce the effects of drift. But if the flow sensor shows drift, a high K_p value will accentuate this drift;

- that systems with high K_p values (ie nervous controllers) are not very stable, they tend to over-react to the input signal. Nevertheless in comparison with simple on/off controllers proportional controllers are capable of reducing deviations from setpoint considerably. If an exact match of the system output and set point is not necessary, a P controller will usually be sufficient.

9.4.2 Integrating elements and proportional controllers

A proportional controller with an integrating element works in the following way. The integrating element continually accumulates (integrates) its input signal. The integrated input signal is multiplied by a factor K_i. With an integrating element we can completely reduce steady state errors.

Π Before reading on, see if you can think of a reason why accumulation of the input signal by the integrating element can reduce the steady state error.

The reason we would give is that as long as the error exists, the integral steadily grows. Therefore, the correcting action gets stronger. This continues until the error has disappeared completely.

A purely integrating controller however, is slow due to the fact that the integral needs time to build up. An integrating element can easily cause overshoot (and even make the controlled system unstable) because the correcting action is getting stronger and stronger.

PI controllers

Normally, an integrating action is combined with a proportional action controller (yielding a PI controller).

9.4.3 Differentiating elements and PID controllers

A differentiating element examines the rate of change of the output of a process. If the output of a process changes, then the differentiating element generates a large controller action. If, on the other hand, a large change occurs in the output of a process but this change occurs over a long time interval, then the differentiating element will generate only a small input signal. The differentiating element multiplies the derivative of its input signal by a constant K_d.

SAQ 9.2

Examine the following graph of output values of a process against time and mark on the stage(s) at which a differentiating element will produce the greatest controller action. (You might find it convenient to draw on graphs plots of dO/dt against t.

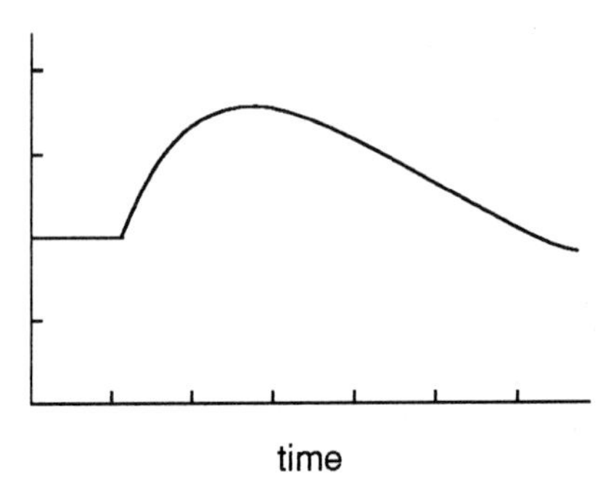

The key feature of a differentiating element is that it detects change. In this way, the controller may act even before the output has changed dramatically (see the example provided in the graph in SAQ 9.2). Thus d differentiating element may be stabilising. A differentiating action makes the controller more nervous. Measurement noise especially has a dramatic effect on the differentiating action.

However this effect can be readily circumvented by pre-filtering the controller input. Differentiating elements are insensitive to slow changes. One can envisage a situation in which the measured output has slowly drifted away from the setpoint value; this would go undetected by differentiating elements.

PD and PID controllers

Differentiating elements can be combined with proportional and/or integrating elements. The controller produced by these combinations are usually referred to as PD and PID controllers.

9.5 State estimation

Consider the following example. In order to maintain optimal process conditions in an ethanol producing fermentation process, the substrate concentration must be kept at a certain low level. This concentration cannot be measured on-line. However, the ethanol concentration can be measured on-line (with limited accuracy). There are two ways of controlling this fermentation. The first is by directly controlling the measured ethanol concentration. The second way is to estimate the substrate concentration by means of a more or less crude model of the process and the on-line ethanol measurements. The estimated substrate concentration can be used to adjust the substrate feed rate (dilution rate) of the process.

There are some important elements in this type of approach. First we need to generate a model that relates the component we can measure (ethanol, in the case above) and the component we wish to control (substrate in the case above). Often the model is quite complex. For example product yield in biological systems is influenced by a very large number of parameters such as growth rate, temperature, pH, available macro and micro nutrients etc. We do not intend to examine the modelling of microbial processes here except to say that to carry out the type of process described above we must generate a model. Modelling of microbial processes is described in detail in the BIOTOL text 'Bioprocess Technology: Modelling and Transport Phenomena'.

importance of modelling

The success of the approach depends on how accurate the model is and how accurate the actual measurements are. We also need to bear in mind what the actual control problem is. You will recall from Chapter 2 in our discussion of regime analysis (Section 2.7) that we can identify rate limiting mechanisms within processes from characteristic times. In other words, we can use characteristic times to define what has to be modelled and what is unnecessary.

We might re-state the problem described above as how do we estimate the substrate concentration as reliably as possible, on the basis of the actual measurements and the available model.

Powerful tools for on-line estimation of process variables are called state observers. State observers are effectively state estimators. State estimation works in the following way. First, a model is used to produce an estimation of the system state, for example in the alcohol, substrate example above we would generate a model which relates the two. Thus:

substrate = f (ethanol)

where f is a function which might be simple or complex.

In principle if we measure alcohol, the estimator can calculate substrate using this model.

The next stage is to experimentally verify the model. This involves measuring the alcohol concentration and using the estimator to calculate (estimate) the substrate concentration. The substrate concentration is also determined experimentally. The differences between the measured values and the estimated values are used to modify the model. A properly 'tuned' state estimator should give an accurate estimate of a process variable.

The advantage of using state estimators is that it enables us to have on-line control of a process variable that cannot be measured on-line. We will compare the advantage of on-line control with off-line measurement in a later section. State estimators can also be used for noise reduction. Note however that when the single goal is to dampen signals (ie reduce noise), many other techniques may be preferred. We cite for example, a statistical routine that averages signals over a period or electronic filters.

9.5.1 Summary of controllers action

Thus far in this chapter we have described the basic layout of control schemes. We met with feedback control, feedforward control, supervisory control and the role of estimators in inferential control. We have also examined what actions controllers may take. These included simple on/off controllers, proportional controls, integrating and differentiating elements. The only part of the control scheme we have, as yet, not examined is the measurement of variables. These we will examine in the next section. Before you move on attempt SAQ 9.3 and SAQ 9.4. A more formal mathematical approach to process control action using transfer functions is provided in Appendix 2.

SAQ 9.3

Answer true or false to each of the following statements.

1) Differentiating elements are capable of detecting small changes providing they occur rapidly.

2) Integrating elements always respond rapidly to changes in output signals.

3) A long time delay in a feedback control system may lead to considerable overshoot.

4) Time between changes in measured values and control action should always be as short as possible.

5) A proportional controller once set up to maintain an output of a process at a setpoint will not require any re-adjustment to ensure the output remains constant.

6) A state estimator allows us to operate on-line control of a variable for which no direct on-line measurements are available.

7) We could call the approach described in 6) 'inferential control'.

SAQ 9.4

It has been established that antibiotics (P) production in a microbial system in continuous culture is related to the growth rate by the following relationship:

$$q_p = \frac{1}{a\,\mu} + \frac{1}{b}\,q_s + m$$

where: q_p is the rate of antibiotic formation in units of moles per gram of biomass per hour; q_s is the rate of substrate used in units of moles per gram of biomass per hour; a, b and m are constants which are sensitive to temperature; a and b are also sensitive to substrate concentration; μ is the specific growth rate of the organism and is sensitive to temperature and substrate concentration.

The product (P) can be measured quite accurately on-line but no such on-line measurement of the substrate can be made.

From this information make a list of the variables that will need to be controlled to produce predictable quantities of P using predictable quantities of substrate. Suggest a strategy for implementing this control.

9.6 Measurement instrumentation

9.6.1 Introduction

definition of on-line and off-line measurement

In order to be able to perform automatic process control, direct process information in the form of measurements is essential. Although it is possible to argue about the definition of on-line and off-line measurements, we will use a straightforward definition. On-line measurements are performed automatically and the results are directly available for control. On-line measurements are therefore usually regular and frequent (or even continuous) in time (note the difference between 'regular' and 'frequent'). Off-line measurements on the other hand require a human interface, for taking samples, performing a (partly) manual analysis and/or for feeding the result into the controller. They are usually much less frequent and are irregular.

It should be noted that the given definition of on-line measurements includes all kinds of automatic sampling and analysis systems. Some of those auto-analysers are relatively slow and expensive to operate. Therefore, these measurements will not be too frequent. The fact that a measurement is carried out on-line is therefore no guarantee that it is fast and frequent.

Π Can we use the information of off-line measurements in a control loop? Think about this before reading on and see if you can devise a scheme. Will it offer good control?

In principle the answer is yes, in practice off-line measurements pose a number of difficulties in producing good process control. Let us examine a process and see if we can highlight the difficulties. Consider a process, for example, involving a bioreactor in which it is necessary to control substrate concentration. Let us assume that there is no direct or inferential method available for on-line measurement. The only method available involves taking a sample and analysing it using off-line equipment. Invariably this will of course introduce a delay time between taking the sample, obtaining the measured value and then adjusting the process. This delay may be from a few minutes to several hours depending on the nature of the variable being measured. Long delay times between sampling and responding are likely to lead to considerable overshoot. Essentially what a long delay time means is that we only know what the process state was some time earlier than when the measurements were made.

The problem is further exacerbated by the rather unreliable nature of manual operations. It is likely that the delay time may vary from measurement to measurement. Thus, due to the manual character of off-line measurements, it is almost impossible to give a dynamical description of the measurement. Even a simple delay time description will vary from measurement to measurement. Therefore, it will be very difficult to use this information in a control loop, even if the measurements are quite frequent.

For efficient automatic process control, one needs on-line measurements (preferably fast measurements). Off-line measurements, on the other hand, are more suited to perform checks, calibrations and to adjust controller settings etc.

9.6.2 On-line measurements

The most common on-line measurements on the fermentation broth are temperature, the pH and the dissolved oxygen concentration. For each of these variables, the most commonly used instruments are given in Table 9.1. Other measurements are gas and liquid flow rates, pressure, stirrer power input, (broth viscosity), broth volume (or mass), foam detection, for which the most common instruments used are also listed in Table 9.1.

Π Table 9.1 is large, so examine it carefully and use the table to make a list of the features which are desirable for on-line sensors.

Below are our ideas about what features are needed to make a good on-line sensor. There are many demands on on-line sensors used in bioprocesses, especially those which involve the growth of micro-organisms or provide conditions in which micro-organisms can grow. In principle on-line sensors must be repeatedly (steam) sterilisable preferably without having to be recalibrated (heat and pressure resistance). Furthermore, they must be mechanically stable in order to withstand high shear forces, sometimes due to solid particles. Sensors should be constructed in such a way that bacterial growth on the membranes etc is prevented as much as possible. Another major demand is of course, that they may not possess any leaks with regard to bacterial or viral infections. The latter also holds for automatic sampling systems. After a sample has been taken, the aseptic circumstances do not have to be maintained. However, some analytical equipment is notorious for the fact that bacterial growth (or precipitation of enzymes etc) may cause clogging. Also, because as far as we are aware, no sensor system exists that allows aseptic replacement, sensors must give a stable and reliable signal over a long period. Usually, only single point calibration of a sensor-in-use is possible. A sensor with linear characteristics, whose signal is zero at zero concentration, is therefore to be preferred. Finally, we want fast measurements. Therefore, the sensor dynamics should be fast in relation to those of the measured variables. This does not mean that all sensors need to be equally fast. For example, the solubility (and, therefore, concentration) of oxygen in culture media is low. Micro-organism use oxygen at a very fast rate and therefore the relative concentrations of oxygen may alter very rapidly. In order to measure and control oxygen concentration we must have fast dynamics. Ethanol however is usually produced in fairly high concentrations and the relative changes in ethanol concentrations are much slower. Thus, for ethanol measurement, much slower sensors are adequate.

The list of features desirable for on-line sensors you could have generated, should include:

- sterilisable (heat and pressure resistant);
- mechanically robust;
- low adhesion of bacterial cells and other fouling species (eg proteins);
- leak proof;
- stable signal over a long period;
- linear characteristics;
- fast dynamics.

variable	measure-ment devices	+ = suitable	+ = accurate	+ = little drift	+ = little delay	remarks
Medium composition						
biomass concentration	spectrophotometer	+/-	-		+	non-linear
	nephelometer	+/-	-		+	signal/dead cells
	fluorometer	+	+		++	
	dielectric permittivity	++			++	living cells only
	dry weight	--	+	+	--	dead cells
	cell count	--		+	--	dead cells
	activity measurement,	--	+	+	--	
	bacterial component anal.	--		+/-	--	
pH	glass electrode	+	+	+	+	fouling
dissolved oxygen	galvanic electrode	++	+	-	+/-	pressure/
	polarographic electrode	++	+	+/-	+/-	fouling
dissolved carbon dioxide	CO_2 electrode	+	+		+/-	
redox potenital	redox electrode	+	+			
volatile medium components	semi-permeable tubing sensor	+		-	-	fouling of membrane
other broth components (ammonium, substrate, etc)	ion selective elctrode	+				
	biosensor:					
	enzyme thermistor	++		-		poorly
	enzyme electrode	++		-		sterilisable/
	affinity sensor	++		-		instable
	gas chromatograph	+	+	+/-*	-	column fouling
	HPLC	+	+	+	-	column fouling
	mass spectrometer	+	+	++	-	
	refractometer	+/-	+	++	+	
	autoanalyser	+	+	-*	**	
	flow injection analyser	+		-*	**	
Gas Composition						
oxygen concentration	paramagnetic analyser					pressure/
	gas chromatograph	++	+	+/-	+	humidity/
	mass spectrometer	++	+	+/-*	-	temp
carbon dioxide concentration	infrared analyser	++	+	+/-	+	pressure/humidity/ temp
	gas chromatograph	++	+	+/-*	-	
	mass spectrometer	++	++	++	-	

Table 9.1 Summary of measurements and their characteristics (continued on the next page). A key to the symbols used is included in the legend.

variable	measure-ment devices	+ = suitable	+ = accurate	+ = little drift	+ = little delay	remarks
other components (CH_4, H_2, ethanol etc)	gas chromatograph	+	+	+/-*	-	
	mass spectrometer	+	+	+	-	
Physical variables						
temperature	temp dependent resistor	++	+	++	++	
	platinum resistor sensor	++	++	++	++	
	thermo couple	+	+	+	+	
	thermometer bulb	--	+	++	+	
gas flow rate	rotameter	--	+	++	++	
	thermal mass flow meter	++	++	++	++	
	differential pressure meter	++	+	++	++	
liquid flow rate	electromagnetic flow meter	++	+	++	++	
	liquid level sensor	++	-	++	++	
	reactor weighing	+		++		
	calibrated pump		+	+/-	++	
pressure	diaphragmatic gauge	+	+	+	+	
liquid volume	liquid level sensor	++	-	++	+	
	reactor weighing	+		++		
foam	contact probe	+		+	+	
power input	hall effect wattmeter	++	--	++	++	frictional losses also meas.
	torsion dynamometer	++	--	++	++	

Table 9.1 continued. * automatic calibration possible, ** strongly dependent on the specific measurement.

Π It would be worthwhile re-checking Table 9.1 to make a list of those instruments which are susceptible to drift, those which have long time delays and those which have the possibility of automated calibration.

Π Can you use information of on-line measurements with a large delay time or slow dynamics in a control loop?

The information derived from on-line measurements with a large delay time is, of course, less valuable than that obtained from fast measurements. However, by including the dynamics of the measuring instrument in the model (which is used for controller development and optimisation), one can often apply this information in the control loop. In general, modelling of the sensor equipment is not necessary if the characteristic time of the measurement is much smaller than that of the dynamics of the measured variable.

You should note that Table 9.1 should not be interpreted too strictly, because most of the given characteristics depend on the exact type of sensor and above all, on individual circumstances. Furthermore, some measurement instruments, eg the auto-analyser, may combine many component measurements at the same time. The characteristic time of these measurements strongly depends on the measured components and methods. You should also note that many on-line devices are in the process of development. We particularly cite the enzyme-based sensors. This is a rapidly developing area and we will not examine these in detail here. For a fuller discussion of enzyme-based sensors, the reader is referred to the BIOTOL text 'Technological Applications of Biocatalysts'. The point we are making is that Table 9.1 provides only an over-view of the devices available for measurement of variables in bioprocesses. It does not pretend to give a comprehensive, nor detailed, list.

9.6.3 Off-line measurements

For off-line measurements special care must be taken in order to obtainrepresentative samples. Large scale bioreactors are notorious for their gradients, ie some components are not equally distributed over the fermentation broth (when mixing and consumption take place on the same time scale). This problem can best be overcome by using several sampling points (eg close to the impeller, close to the influent inlet, in a dead-corner, etc). The representative sample is then a mixture of the individual samples. A simpler method is to take the sample slowly at an average spot, allowing some time to have the peak values mediated.

Π Write down a reason why it is this less important to take care to obtain representative samples for on-line measurements?

We believe that the reason is due to the fact that on-line measurements are usually quite frequent, thus possible artifacts will be mediate. This does not hold for the off-line analysis where measurements are usually taken infrequently. The exact location of the on line sensors should, however, be well chosen otherwise a systematic error may result.

Another problem with off-line measurement is that of the conservation of the sample. For instance, the dissolved oxygen concentration cannot be reliably analysed by sampling, because the oxygen concentration will be changed either by consumption after the sample has been taken or by oxygen from the air that dissolves in the sample. An equal problem is apparent if a low substrate concentration is to be analysed (eg from a bacterial culture growing under substrate limitation). If there are still some bacteria in the sample, within seconds, the remaining substrate can be exhausted. Therefore, the bacterial activity in the sample must be repressed as fast as possible. In general, there are two ways to accomplish this:

- suck the sample through a bacterial filter;
- use a sample vessel which is either very cold or contains a chemical substance (eg acid) that immediately stops all bacterial activity.

Note, however, that the use of acid may cause difficulties in enzymatic, and other types of analysis.

Normally, the solid particles (mainly biomass) are separated from the broth by means of centrifugation. Liquid (the supernatant) and solid phase (the pellet) can then be analysed separately. Numerous analytical methods are applied. Often, a specific solution must be found for each situation: an analysis that performs satisfactorily for one medium may be inadequate for another type of broth. Finally, we should mention that, due to the manual character of many off-line measurements, errors in the measurement procedure are more likely, while at the same time these errors are more difficult to track down. Even the use of duplicate measurements cannot always solve this problem, as these duplicates are often analysed in the same sequence.

SAQ 9.5

Choose the appropriate words from the list provided to complete each of the following statements.

1) Biosensors based on enzymes are difficult to use for on-line measurement because they are often [ ].

2) Biosensors based on enzymes are desirable to use for on-line measurement because they are [ ].

3) A common problem with using HPLC analysis measurement of medium composition is that of [ ].

4) The basic problem with using spectrophotometry to measure biomass concentration is that it produces a [ ] signal.

Word list: column fouling, inaccurate, non-linear, specific, unstable.

9.6.4 Consistency - checks on measurements

Whether process control is based on direct measurements or on state estimates, in both cases the measurements should be reliable. For some processes, it may therefore be advisable to perform some measurements more than once, in order to improve the reliability of the measurements (eg by using more than one sensor or analysing duplicates). This kind of redundant information can be used to check whether a measurement is correct. In this section, however, we will concentrate on another type of check on the measurements, based on the law of conservation for elements (or energy). Still, redundancy is the key.

Let us assume the following conversion takes place in the system (only the overall conversion is stated):

glucose + NH_3 + oxygen $\rightarrow$ biomass + water + ethanol + carbon dioxide

We can write the stoichiometry as:

$$r_s.CH_2O + r_n.NH_3 + r_o.O_2 \rightarrow r_x.CH_{1.8}O_{0.5}N_{0.2} + r_h.H_2O + r_p.CH_3O_{0.5} + r_c.CO_2$$

where:

r = conversion rates

r_s, r_n and r_o are negative since these components are being consumed.

Note that we have described the biomass as containing the elements C, H, O and N in the proportions of 1:1.8:0.5:0.2.

Since carbon cannot be created or destroyed during the conversion we can write a carbon balance where:

$r_s + r_x + r_p + r_c = 0$ (remember r_s is negative)

or

$- r_s = r_x + r_p + r_c$

Note that the derivation of elemental balances in bioconversions are described in detail in the BIOTOL text 'Bioprocess Technology: Modelling and Transport Phenomena'.

Similar equations can be written for the other elements. Thus for hydrogen.

$(-2r_s) + (-3r_n) = 1.8r_x + 2r_h + 3r_p$

and for oxygen:

$(-r_s) + (-2r_o) = 0.5r_x + r_h + 0.5r_p + 2r_c$

Π Write down the balance for nitrogen.

Your answer should have been $-r_n = 0.2r_x$ or $0.2r_x + r_n = 0$

In this system we have 7 conversion rates (ie r_s, r_n, r_o, r_x, r_p and r_c). We also have 4 equations from the elemental balances which relate the rates to each other. Thus, in principle, if we measure 3 of the rates, the remainder of the rate values can be calculated using these balance equations. Such a system is said to be determined.

However, if more than 3 rates are measured, the possibility arises to check the measurement set, because the given balances may not be violated. The system is said to be over-determined.

Π Let us try an example. We will use the bioconversion system described above in which glucose is converted to biomass, ethanol and carbon dioxide.

In our system four rates have been determined. These are:

- the rate of substrate consumption $r_s = -61 \times 10^{-2}$ C-mol $l^{-1}h^{-1}$;
- the rate of production of biomass $r_x = 40.6 \times 10^{-2}$ C-mol $l^{-1}h^{-1}$
- the rate of carbon dioxide production $r_c = 20.4 \times 10^{-2}$ mol $l^{-1}h^{-1}$
- the rate of oxygen consumption $r_o = -20.5 \times 10^{-2}$ mol $l^{-1}h^{-1}$.

Our task is to use the oxygen consumption rate to determine if there are errors in our measurements.

We will begin by explaining two of the values. The substrate consumption rate (r_s) has been given in units of moles of substrate carbon per litre per hour. Likewise the rate of biomass production (r_x) has also been given in terms of moles of biomass carbon per litre per hour. The important thing to bear in mind is that we must use the same units for each term otherwise we can get into great difficulty.

Essentially what the question is asking us to do is to use r_s, r_x and r_c to calculate r_o on the basis of conservation balances and to see if the calculated r_o is the same as the measured r_o. If they are different, then we know that there must either be an error in the stoichiometry we have used to describe the conversion or there is an error in one, or more, of the measured r values.

Let us do the necessary calculations. The relationships we have are $r_s = r_x + r_p + r_c$

$$(-2r_s) + (-3r_n) = 1.8\ r_x + 2\ r_h + 3\ r_p$$

$$-r_s + (-2r_o) = 0.5\ r_x + r_h + 0.5\ r_p + 2\ r_c$$

$$-r_n = 0.2\ r_x$$

First we can calculate $r_n = -0.2\ r_x = -8.12 \times 10^{-2}$ mol $l^{-1}h^{-1}$

We can now calculate r_p from $-r_s = r_x + r_p + r_c$

$$-(-61 \times 10^{-2}) = 40.6 \times 10^{-2} + r_p + 20.4 \times 10^{-2}$$

$$r_p = 0 \text{ mol } l^{-1}h^{-1}$$

We can now calculate r_h from $(-2r_s) + (-3r_n) = (1.8r_x) + (2r_h) + (3r_p)$

$$(-2 \times -61 \times 10^{-2}) + (-3 \times -8.12 \times 10^{-2}) = (1.8 \times 40.6 \times 10^{-2}) + 2r_h + 3 \times 0$$

$$73.28 \times 10^{-2} = 2r_h$$

Thus:

$$r_h = 36.64 \times 10^{-2} \text{ mol } l^{-1}h^{-1}$$

Finally we can calculate:

r_o from $(-r_s) + (-2r_o) = 0.5r_x + r_h + 0.5r_p + 2r_c$

$61 \times 10^{-2} + (-2r_o) = 20.3 \times 10^{-2} + 36.64 \times 10^{-2} + 40.8 \times 10^{-2}$

$-2r_o = 36.74 \times 10^{-2}$

Thus:

$r_o = -18.37 \times 10^{-2}$ mol $l^{-1}h^{-1}$

This value is different from the actual measured value of r_o (= -20.5×10^{-2} mol $l^{-1}h^{-1}$) so there is some error in at least one of the measured rates (ie the error in r_o is $-20.5 \times 10^{-2} - (-18.37 \times 10^{-2}) = -2.13 \times 10^{-2}$ mol $l^{-1}h^{-1}$).

Π How might we attempt to determine which measurement(s) was producing the error?

One way would be to measure another one the rates eg r_n. This would enable us to use a variety of combinations of values to calculate the remainder (eg we could use r_s, r_c and r_n or r_s, r_o and r_n etc). In this way, we would be able to determine which measured r values were consistent with each other, which in turn would enable us to determine which measured r value was erroneous. Of course if more than one r value is measured inaccurately, life can become rather difficult!

Returning to our example above if we measured r_n and found it not to be -8.12×10^{-2} mol $l^{-1}h^{-1}$ we would immediately be suspicious that the measured r_x value was inaccurate.

It would go beyond the scope of this text to go into the use of statistics. Nevertheless it is important to note that a statistical treatment will not only provide us with information on whether or not the measurement set is reliable, but it also yields an improved set of data. It also gives us a measure of confidence we can apply as to the accuracy and precision of measured values.

9.7 Optimisation

We conclude this chapter by a brief consideration of process optimisation. Although strictly speaking optimisation does not belong to the area of process control, the control engineer will in most cases be charged with the problem.

We can broadly divide the strategy for optimisation into two types:

- based on laboratory experiments or previous process runs;
- based on a model of the process.

We will not consider optimisation based on experimental work any further here except to say that clearly the intention is to generate practical evidence that enables the selection of optimal controller set points etc. For example such evidence will enable setting the process temperature, substrate feed rate etc. However sometimes, we may have to develop a process strategy (preferably optimal) using a process model. The success of this, of course, depends upon the quality of the model. If the model does not exactly match the system then the calculated working and set points may be in error. Nevertheless, although not ideal, such models do, at least, generate serviceable starting points.

Optimisation based on models and experiments can be divided into two categories:

- the optimisation of steady state systems;
- the optimisation of dynamic systems.

In the first category we might cite a continuous fermentation system. In the second category we could use batch-fed cultures. Thus in this second category we could be concerned with optimising the feeding strategy. To optimise such a process, one basically needs to simulate the whole process from beginning (inoculation) to end (harvest). By using different variables (eg substrate concentration, times of feeding etc) and repeating the simulation it becomes possible to determine the optimum conditions for the process. Essentially what is needed is to determine the complete time-path of the variable throughout the process. A good example is the optimisation of the feeding strategy of a fed batch culture. In principle we need to find the optimal influent substrate concentration at each moment in time. As there are an infinite number of moments in time, the optimisation problem is of an infinite order. Note that if we have two interdependent variables (eg substrate concentration and product yield) then we end up with a problem of $(\text{infinite})^2$! Mathematically there are ways of solving such problems (eg the Principle of Pontryagin) but these require considerable mathematical expertise for their implementation.

An alternative, and more accessible method is to use dynamic programming. In this, discrete time values are used for the input. This reduces the order of the optimisation problems. Then repeated simulations are carried out using a range of inputs. This enables identification of the optimum inputs for the process.

Often dynamic problems are translated to static ones by the engineers' experience and intuition. For example, instead of finding the complete optimal time-path for a fed-batch culture, one may simply demand a maximal growth rate during the first phase of the process and a maximal production rate during the second phase. This will yield the required state setpoints which can be attained by specific dynamic input signals.

Summary and objectives

This chapter has provided an overview of process control. In it, we first explored the basic control schemes that are available including feedback, feedforward, inferential and cascade control systems. We went on to explore the operation of basic controllers including various types of proportional controllers. We also examined the range of instrumentation available for on-line and off-line measurements, considered the merits of on-line measurement and introduced a method to check for consistency. In the final section we briefly considered process optimisation.

Now that you have completed this chapter you should be able to:

- describe the range of control schemes which exist;
- demonstrate an understanding of how various types of controllers work;
- understand and use a variety of technical and jargon terms used in control technology;
- explain how overshoot might occur with a feedback system with a long delay time;
- establish what variables will need controlling;
- identify the features of measurement devices which influence their use as on-line devices;
- use conservation principles, especially of elements, to check for consistency in measurements;
- suggest sensitive input signals for a specific control problem;
- list sensor requirements, including sensor speed/delay time, with regard to characteristic times of the process.

Responses to SAQs

Responses to Chapter 2 SAQs

2.1

1) True. The figure in the question shows the plots for the three different scales of operation giving the same relationship between product yield and P/V up to about P/V of 1 kWm^{-3}. Above that value of P/V, the plots diverge.

2) False. Although the pilot plant data accurately represent production scale operations up to about P/V = 2 kWm^{-3}, at higher P/V values, the plots diverge.

3) True. From the laboratory scale operation we would predict we would need higher P/V ratios than would actually be required to achieve a particular yield.

4) False. The opposite is in fact true - the yield would be higher than we would predict from laboratory scale operation for P/V values above 1 kWm^{-3}.

2.2

1) Coalescing. Our reasoning is as follows:

$$\text{In general } k_La = c\left(\frac{P_g}{V}\right)^a (v_s)^b$$

We have been told that all parameters except V_s have remained constant. Thus:

k_La is proportional to $(v_s)^b$ for the two conditions.

Thus if we let $k_La = y$ for the first set of conditions:

$y \alpha (x)^b$. Note we are using α to represent proportionality.

For the second set of conditions:

$$\frac{y}{2} \alpha \left(\frac{x}{4}\right)^b$$

$$\text{or } y \alpha 2\left(\frac{x}{4}\right)^b$$

$$\text{Thus } (x)^b = 2\left(\frac{x}{4}\right)^b \rightarrow b = 0.5$$

This is the exponent typical of a coalescing medium used on a large scale.

2) The most likely explanation is that the original medium is highly viscous and that the growing organisms has broken down the polymer which results in a viscosity decrease.

Remember that $k_La = c\left(\frac{P_g}{V}\right)^a (v_s)^a (\eta)^{-0.86}$ (Equation 2.7)

for viscous broths. Thus if η is decreased, then k_La will increase despite there being no changes in the power input, superficial gas velocity etc.

2.3

1) Typical value for α is about 0.4.

2) You should have chosen the following boxes.

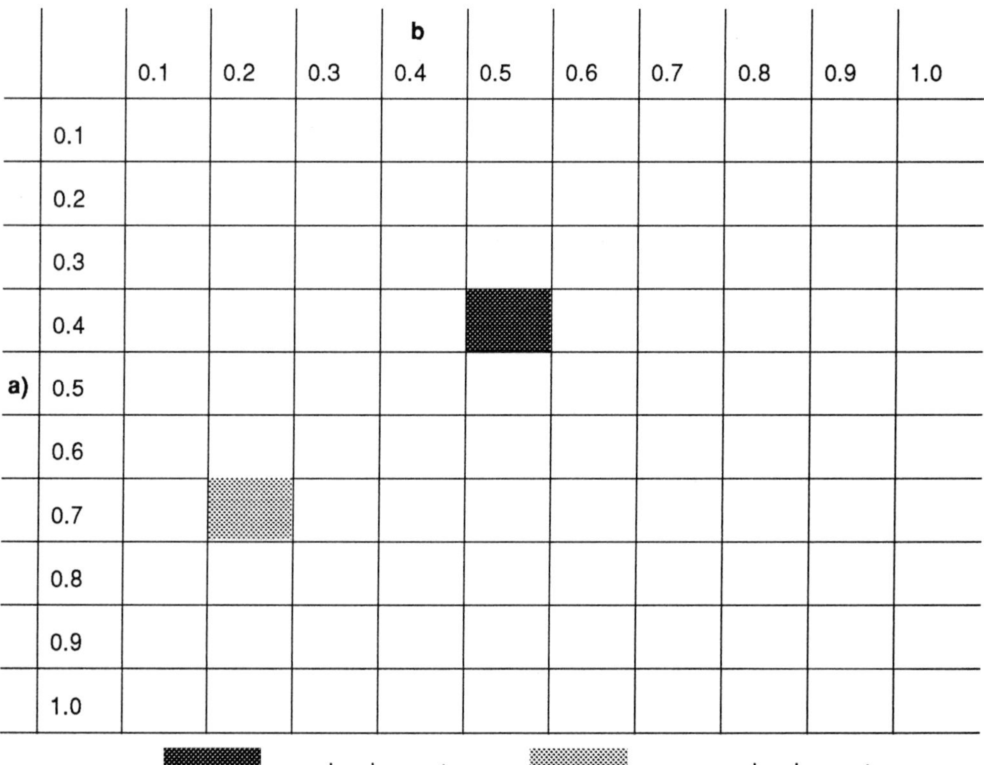

a) \ b	0.1	0.2	0.3	0.4	0.5	0.6	0.7	0.8	0.9	1.0
0.1										
0.2										
0.3										
0.4										
0.5										
0.6										
0.7										
0.8										
0.9										
1.0										

Note that a system that is non-coalescing on a small scale tends to become coalescing at large scale. So it is reasonable to assume that the values of a and b for a large scale (non-) coalescing system are a = 0.4, b = 0.5.

3) 2 kWm^{-3} see Table 2.6 and text.

4) In experiments with CFSTR, calculated mixing times (t_m) are often shorter than the measured t_m values (see Table 2.5 and associated text).

2.4

1) In this problem we have 5 parameters and 3 different units (kg, m, s). Thus applying the Buckingham II theorem, we would anticipate a minimum of 2 dimensionless numbers (p = n - m = 5-3).

2) We begin with the given equation:

$$C_x = C_o^{\alpha} \,.\, D^{\beta} \,.\, v^{\gamma} \,.\, L^{\delta}$$

Substituting in the units:

$$\left[\frac{kg}{m^3}\right]^1 = \left[\frac{kg}{m^3}\right]^{\alpha} \,.\, \left[\frac{m^2}{s}\right]^{\beta} \,.\, \left[\frac{m}{s}\right]^{\gamma} \,.\, [m]^{\delta}$$

We can now examine each unit in turn.

For kg: $1 = \alpha$

For s: $0 = -\beta - \gamma$ Thus $\beta = -\gamma$

For m: $-3 = -3\alpha + 2\beta + 1\gamma + \delta$

But $\alpha = 1$ and $\beta = -\gamma$

Thus $-3 = -3 + 2\beta - 1\beta + \delta$

Thus $\beta = -\delta$

Thus $\alpha = 1$; $\beta = -\delta; \gamma = \delta$

Substituting into: $C_x = C_o^{\alpha} \,.\, D^{\beta} \,.\, v^{\gamma} \,.\, L^{\delta}$

$$C_x = C_o^{1} \,.\, D^{-\delta} \,.\, v^{\delta} \,.\, L^{\delta}$$

or grouped into dimensionless numbers

$$\frac{C_x}{C_o} = \left[\frac{vL}{D}\right]^{\delta}$$

Thus our two dimensional numbers are $\frac{C_x}{C_o}$ and $\frac{vL}{D}$

3) Careful examination of Table 2.9 shows that:
$\frac{vL}{D}$ is the Péclet number (you should have found this fairly easy since we are clearly dealing with mass transfer).

What this analysis shows is that the ratio of substrate concentration at the outflow to that of the inflow is a function of the Péclet number.

Thus we can try to maintain the Péclet number constant as our scaling criterion.

For this we can write:

$$\frac{C_x}{C_o} = \left[\frac{vL}{D}\right]^{\delta} = [Pe]^{\delta}$$

Since D will be the same at all scales of operation then if the model value L is reduced compared to the prototype, v must be increased in direct proportion in order to retain a constant $\frac{C_x}{C_o}$ ratio.

2.5

1) 29°C. We approached this in the following way. The first stage is to plot the temperature profile in the model pipe in a dimensionless form. We can convert the temperature into a dimensionless form by using the ratio of the inlet temperature to the actual temperature at different points along the pipe (ie $\frac{T_x}{T_o}$). Likewise we can convert the distance (x) along the pipe into a dimensionless form by describing it as a fraction of the total length of the pipe ($\frac{X}{L}$). Thus for the values given in the question.

x (cm)	$\frac{X}{L}$	T_x (°C)	$\frac{T_x}{T_o}$
0	0	60	1
25	0.25	50	0.83
50	0.5	42	0.7
75	0.75	35	0.58
100	1	29	0.48

Thus, in dimensionless form:

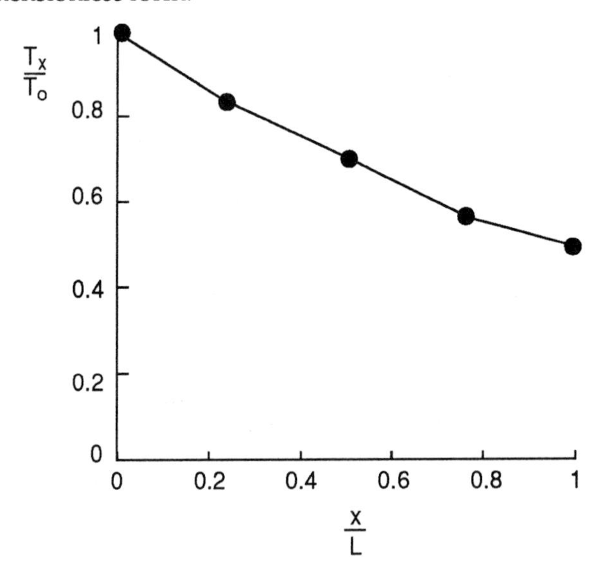

The velocity of the water in the model is 0.1 ms^{-1}. Since the length of the pipe is 1m, the water is in the pipe for $\frac{1}{0.1} = 10s$.

Similarly, for the prototype, the velocity of the water is 10 ms^{-1} and the length of the pipe is 100m, thus the water is in the pipe for $\frac{100}{10}$ = 10s. Thus this is the same for both systems. Thus the dimensionless temperature profile for the model should be the same as the dimensionless profile for the prototype.

We were asked for the temperature at which the water would leave the prototype, that is when x = L. Thus $\frac{x}{L}$ = 1. From the profile plotted above, when $\frac{x}{L}$ = 1, then $\frac{T_x}{T_o}$ = 0.48. Since T_o = 60°C, T_x = 29°C.

2) To plot the actual (dimensioned) temperature profiles in the model and prototypes we need to plot Tx against x. The values for the model were given in the question. For the prototype, we can use the values provided by the graph plotted in the answer to 1) using T_o = 60°C and L = 100m. Essentially what we need to do is to read the co-ordinates of the points on the dimensionless graph and multiply those values on the vertical axis by 60°C to get Tx and those values on the horizontal axis by 100 m to get the distance x for the prototype.

Graphically therefore we get:

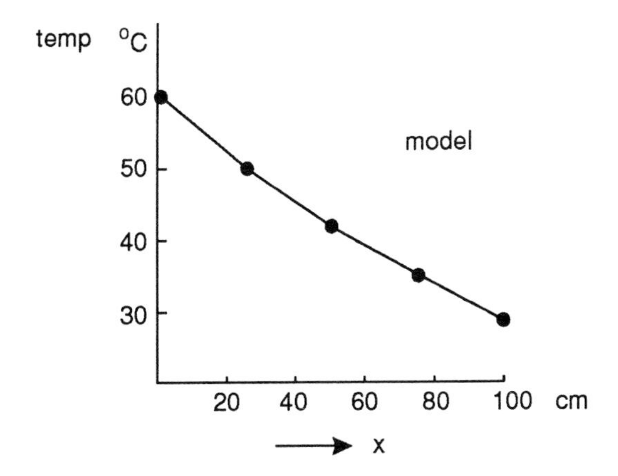

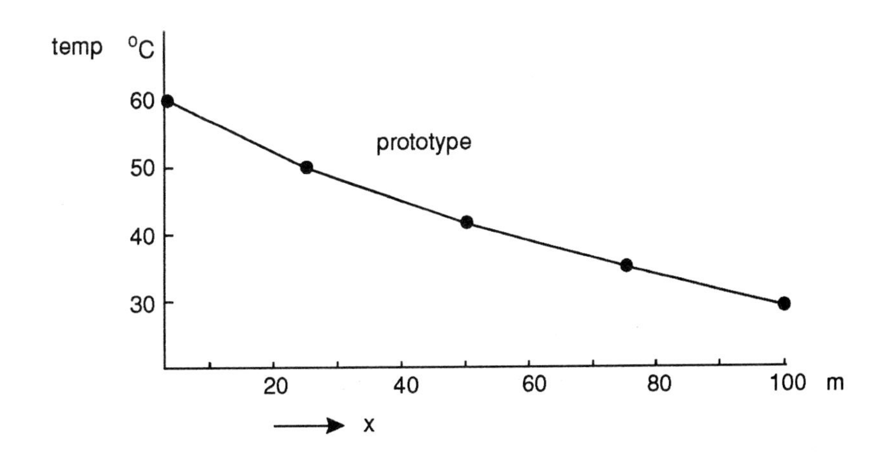

In many ways this problem was rather simple and you may have come to the right conclusion by intuition. Nevertheless it will have demonstrated the principles behind using dimensionless parameters to produce profiles and how we can use model profiles to predict prototype profiles.

2.6

1) Mixed

2) Change

3) We were looking for you to use the word 'characteristic' as this is the word we prefer but in the literature 'process' and 'relaxation' are also used.

4) Transport phenomena are dependent on reactor type, conversion phenomena are not.

2.7

1) It will be doubled.

For an individual bubble $t_{OT,b} = \frac{H}{k_L a}$ (Equation 2.25)

Since k_La is constant and H is doubled, $t_{OT,b}$ will also double.

2) t_{OT} will be halved.

Since $t_{OT} = \frac{1}{k_L a}$ (Equation 2.17)

k_L is constant and a is doubled, then t_{OT} is halved. This is, of course, a common sense answer, if we double the surface area of the bubbles, we would anticipate that oxygen could be transferred at twice the original rate and that this would half the value of t_{OT}.

3) $\phi_p = 0.00075\ m^3s^{-1}$

Since $\phi_p = 0.75 . N . D^3$ (Equation 2.19)

and $N = 1s^{-1}$ and $D = 0.1$ m

4) $t_c = 1333s$

Since $t_c = \frac{V}{\phi_p}$ (Equation 2.18)

then: $t_c = \frac{1}{0.00075}\ s = 1333\ s$

5) $t_m = 5333s$ (approximate)

Since: $t_m = 4 \times t_c$ (see text just prior to Equation 2.18)

6) $v_{lc} = 0.1\ ms^{-1}$

Since $t_c = \frac{H}{v_{lc}}$ (Equation 2.21) and H = 1m, t_c = 10s

2.8

1) 40 s

Since C_{O2} = x mol m^{-3} and K_{O2} is $\frac{x}{1000}$ mol m^{-3} we are using concentrations much higher than the saturation constant. Thus we are dealing with zero order type of kinetics. Thus:

$$t_{oc} = C_{O2}/r_{O2}^{max} = x/\frac{x}{40} = 40\ s$$

2) There will be oxygen limitation. Since transfer of oxygen has a longer characteristic time (t_{OT} = 15s) than the characteristic time for consumption (t_{oc} = 10s), this means that the oxygen is being transferred in slower than it is being used. Thus oxygen will become limiting.

3) Oxygen will not be limiting

Since $t_{OT} = \frac{1}{k_La}$ (Equation 2.17), then doubling k_La will half t_{OT}

Thus $t_{OT} = \frac{15}{2} = 7.5\ s$

This is now shorter than t_{oc} indicating oxygen can be transferred in faster than it can be used.

4) 5000 s. Steady state specific growth rate (μ) = dilution rate (D). Dilution rate = $\frac{\text{flow rate}}{\text{volume}} = \frac{F}{V}$

But $F = \frac{V}{5000}$ (see question), thus $D = \frac{V}{5000V} = \frac{1}{5000}\ s^{-1}$

Thus $\mu = \frac{1}{5000}\ s^{-1}$

But $t_g = \frac{1}{\mu}$ (Equation 2.30) thus t_G = 5000 s.

Responses to Chapter 3 SAQs

3.1

This question basically required you to calculate if oxygen or substrate was limiting.

For oxygen we can use a regime analysis to determine the characteristic times for oxygen transfer and oxygen consumption. The characteristic time for oxygen transfer in the system is given by $t_{OT} = \frac{1}{k_L a}$ (Equation 2.17)

$= \frac{1}{0.04} = 25$ s (since $k_L a = 0.04\ s^{-1}$)

The characteristic time for oxygen consumption is given by:

$t_{oc} = C_{O2}/r_{O2}^{max}$ (Equation 2.28a), since $C_{O2} >> K_{O2}$

$= 10/0.1$

Thus $t_{oc} = 100s$.

Since $t_{oc} >> t_{OT}$ then oxygen availability is not rate limiting.

For substrate:

The rate of substrate input per unit volume = concentration of substrate x dilution rate/unit volume of vessel

$= 1 \times \frac{1}{3600}$ mol $m^{-3}\ s^{-1}$ = 0.00028 mol $m^{-3}\ s^{-1}$

since $C_{s0} = 1$ mol m^{-3} and $D = 1h^{-1} = \frac{1}{3600}\ s^{-1}$.

The maximum rate of substrate use $r_s^{max} = 0.001$ mol $m^{-3}\ s^{-1}$

Thus the rate of substrate is rate limiting.

The analysis suggests that if we wish to improve the performance of the process we should increase the rate of substrate input. A good starting point would be to increase the concentration of the substrate in the input. If this was increased by a factor of $\frac{0.001}{0.00028}$ mol $m^{-3}\ s^{-1}$ then substrate input should balance substrate consumption. This of course would then lead to a change in r_{O2}^{max} so we would then need to determine whether or not oxygen availability had become rate limiting under this new substrate regime.

What this question will have shown you is how to apply a regime analysis to identify the rate limiting mechanism and how upon modifying a parameter (in this case substrate input) another mechanism may become rate limiting. You should however, realise that this question is a gross simplification of the real situation. For example we have neglected the effects of changing input substrate concentration on other parameters such as the viscosity of the medium, the changes in heat output and so on. Nevertheless you should have identified an underpinning strategy namely:

- conduct a regime analysis on the production process;
- identify the rate limiting mechanism;
- use a scale down simulation to modify the process parameters to overcome the rate limiting mechanism, keeping other parameters the same;
- examine the consequences of the changes to the parameter(s) governing the rate limiting mechanism to identify the new rate limiting mechanism;
- modify a parameter which influences this new rate limiting mechanism;
- examine the consequences of these second round of changes.

By repeating this process of sequentially identifying rate limiting mechanisms and modifying parameters, we will eventually optimise the process. Then we can translate these changes implemented at the laboratory scale to their application at the production scale. You must, however, realise that there are inevitably some built in constraints at the production scale including the cost and availability of equipment, running costs etc.

3.2

1) No - Since the characteristic consumption time for oxygen is longer than the characteristic oxygen transfer, oxygen limitation should not be a problem.

2) No - Since the characteristics settling time (8.6×10^4 s) is much longer than the liquid mixing time (60 - 118 s).

3) The answer to this is not quite so obvious. Since the liquid mixing time and oxygen transfer times are relatively similar we might anticipate that gradients of oxygen might be established. But the rate of consumption of oxygen is quite low relative to the rate of oxygen transfer therefore the production of gradients should not cause a problem.

3.3

1) Yes. Since the time needed to transfer oxygen to the liquid phase is much longer than that for oxygen consumption, oxygen limitation will occur. The vessel will, in effect, become depleted of oxygen.

2) Yes. The depletion of oxygen combined with the long mixing time, together with a zone of high oxygen transfer rate for example near the sparger, will result in oxygen gradients.

3) Ethanol. Since the production of ethanol is favoured by low oxygen concentrations, we might anticipate that ethanol production rather than the oxidative production of CO_2 will occur. This will of course depend upon the availability of substrate.

3.4

1a) Mixed regime. Since liquid mixing time (t_m) and oxygen transfer time (t_{OT}) are similar, then these two regimes both rule the performance of the reactor.

1b) No. Since characteristic times for the consumption of oxygen (t_{OC}) is longer than that the transfer of oxygen (t_{OT}), oxygen will not become depleted.

2) We would recommend you draw this out graphically. Thus:

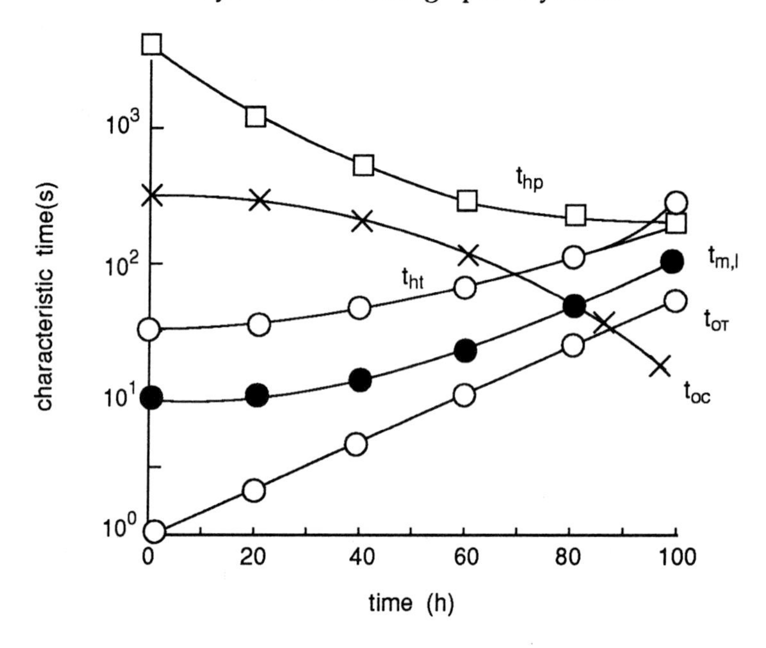

We have made this quite complex. Consider first of all oxygen. At times up to about 90 h, t_{OT} is shorter than t_{OC}, thus oxygen is not rate limiting. At about 90 hours however, t_{OT} becomes longer than t_{OC} thus we must conclude that oxygen becomes limiting at this time. Note also that as the culture progresses $t_{m,l}$ values become progressively similar to t_{OT} and t_{OC} and we must conclude that there is an increasing chance that oxygen gradients will be created as the batch time increases.

Note also that in the early stages of the culture the characteristic time for heat production (t_{hp}) is long, but becomes shorter as the time of culture becomes extended. In contrast, heat removal (heat transport t_{ht} becomes longer) such that t_{hp} and t_{hT} are more or less the same by 100 hours. Thus heat generation and removal are more or less in balance. But t_{hp} and t_{ht} are similar to the length of the mixing time ($t_{m,l}$) thus there is an increasing chance that temperature gradients will be established.

Thus we have described a system in which the reactor was well mixed with no oxygen limitations or over-heating in the earlier stages but with time the reactor becomes progressively less well mixed and both oxygen limitation and the threat of localised over-heating (temperature flucculations) become increasingly important with time.

Responses to Chapter 4 SAQs

4.1 1 g l^{-1}. This was in some ways a rather trivial question. Nonetheless it will have tested whether or not you have understood the previous section.

When we reach a steady state there is no further change in concentration of the compound. Thus:

$$\frac{VdC_A}{dt} = 0 = F(C_{Ai} - C_{Al}) + R_A . V$$

But $R_A = 0$ (given in the question).

Thus $F(C_{Ai} - C_{Al}) = 0$

Therefore $C_{Ai} = C_{Al} = 1 \text{ g } l^{-1}$

4.2 $R_{MS} = 0.1$ mol biomass C l^{-1} h^{-1}

1) $R_{MS} = \mu C_{Ml}$ and $\mu = 1$ h^{-1} and $C_{Ml} = 0.1$ mol biomass C l^{-1}

Thus $R_{MS} = 1 \times 0.1$ mol biomass C l^{-1} h^{-1}

We have assumed that R_{Ms} is proportional to biomass concentration.

2) R_{Ms} will be doubled = 0.2 mol biomass C l^{-1} h^{-1}

4.3 The rate of substrate consumption would be doubled.

If the microbial cell yield (Y_{MSo}) is halved, it means that we need to use twice as much substrate to produce the same amount of cells (see Equation 4.12). Some people can never remember whether the substrate consumption rate is proportional to or inversely proportional to yield. It is fairly easy to remember if you think in the following way. Increased efficiency (ie increased yield) means less substrate has to be used to make the same amount of product. Thus rate of consumption is inversely proportional to yield.

4.4 Using L to denote linear measurements, M = mass, t = time. Thus:

$$V . \frac{d(C_{Sl})}{dt} = F . (C_{Si} - C_{Sl}) - \mu . C_{Ml} . V . \frac{1}{Y_{MSo}} - r_P . C_{Ml} . V . \frac{1}{Y_{PS}}$$

$$\frac{L^3 . M_S}{L^3 . t^1} [=] \frac{L^3 M_S}{t^1 L^3} [=] \frac{1}{t} . \frac{M_M}{L^3} . L^3 . \frac{1}{M_M . M_S^{-1}} [=] \frac{M_P}{M_M^1 . t^1} . \frac{M_M}{L^3} . L^3 . \frac{1}{M_P . M_S^{-1}}$$

Each term in Equation 4.17 has units of $M_S t^{-1}$. Therefore, the terms in Equation 4.17 are consistent.

4.5

1) $\mu = 0h^{-1}$. Since $\mu = D$ and $D = 0\ h^{-1}$ then μ must also = 0.

2) We would get 'washout'. Since the micro-organisms cannot keep up with the dilution rate, they are washed out of the reactor.

3) With no product formation Equation 4.27 reduces to $C_{Si} - C_{Sl} = \frac{C_{Ml}}{Y_{MSo}}$ and therefore

$$Y_{MSo} = \frac{C_{Ml}}{C_{Si} - C_{Sl}}$$

Since the amount of substrate consumed is now only used for microbial growth.

4) $C_{Si} - C_{Sl} = \frac{C_{Pl}}{Y_{PS}}$ since the amount of substrate consumed is now only used for product formation.

4.6

1) 2.4 mol biomass C m^{-3}

Since $C_{Ml} = (C_{Si} - c_{Sl})\ Y_{MSo} = (5 - 0.2)\ 0.5 = 2.4$ mol biomass C m^{-3}

2a) 1 mol product C $m^{-3}\ h^{-1}$

$$\text{Since } C_{Pl} = C_{Ml}\frac{k_{Pn}}{D} = \frac{10 \times 0.1}{1} = 1 \text{ mol product C m}^{-3}\text{ h}^{-1}$$

2b) 2 mol product C $m^{-3}\ h^{-1}$

$$\text{Since } C_{Pl} = C_{Ml}\frac{k_{Pn}}{D} = \frac{10 \times 0.1}{0.5} = 2 \text{ mol product C m}^{-3}\text{ h}^{-1}$$

a) and b) are common sense since in b) by halving D we increase the residence time of cells in the reactor. Thus they have twice as much time to make the product and thus twice as much product will accumulate in the vessel.

4.7

1) Since the microbial cell mass production rate in a CFSTR is DC_{Ml} (see Equation 4.24), then $R_M = DC_{Ml}$. The specific microbial cell mass production rate, r_M is = $\frac{R_M}{C_{Ml}} = \frac{DC_{Ml}}{C_{Ml}} = D$. Remember that at steady state $D = \mu$.

2) Likewise, the rate of formation of microbial product mass R_p in a CFSTR is $D\ C_{Pl}$ (Equation 4.25). The specific rate of formation of microbial product mass in a CFSTR, $r_P = D \cdot \frac{C_{Pl}}{C_{Ml}}$

4.8

0.09g product C (g substrate C)

The specific rate of substrate consumption = specific rate of substrate consumption for growth + specific rate of substrate consumption for product formation.

Thus:

$$r_s = \frac{\mu}{Y_{MSo}} + \frac{r_P}{Y_{PS}}$$ (from combining Equation 4.23 and 4.34)

But, at steady state $\mu = D$

Thus:

$$12 = \frac{0.5}{0.5} + \frac{1}{Y_{PS}}$$

$$11 = \frac{1}{Y_{PS}}$$ Thus $Y_{PS} = 0.09$ g product C (g substrate C)$^{-1}$

4.9

1) 0.917 h^{-1}

Since the point of maximum output rate (D_{max}) is:

$$= \mu_{max}\left[1 - \left[\frac{K_s}{K_s + C_{Si}}\right]^{\frac{1}{2}}\right]$$ (see Equation 4.39) then,

$$= 1\left[1 - \left[\frac{1}{1 + 143}\right]^{\frac{1}{2}}\right] = 1\left[1 - \frac{1}{12}\right] = 0.917h^{-1}$$

2) 9.17g C m^{-3} h^{-1}

Since $R_P = D\,C_{P1}$ (Equation 4.36b), at maximum $R_P = D_{max}\,C_{P1} = 0.917 \times 10$ g C m^{-3} h^{-1}

3) 36.68 g C m^{-3} h^{-1}

4.10

The best way to determine this is to plot $\frac{1}{Y_{MSo}}$ against $\frac{1}{D}$

Thus:

D (h^{-1})	$\frac{1}{D}$ (h)	Y_{MSo} (GC(gC)$^{-1}$)	$\frac{1}{Y_{MSo}}$ (GC(gC)$^{-1}$)
1	1	0.495	2.02
0.5	2	0.490	2.04
0.25	4	0.480	2.08
0.12	8.3	0.460	2.17
0.06	16.7	0.428	2.31

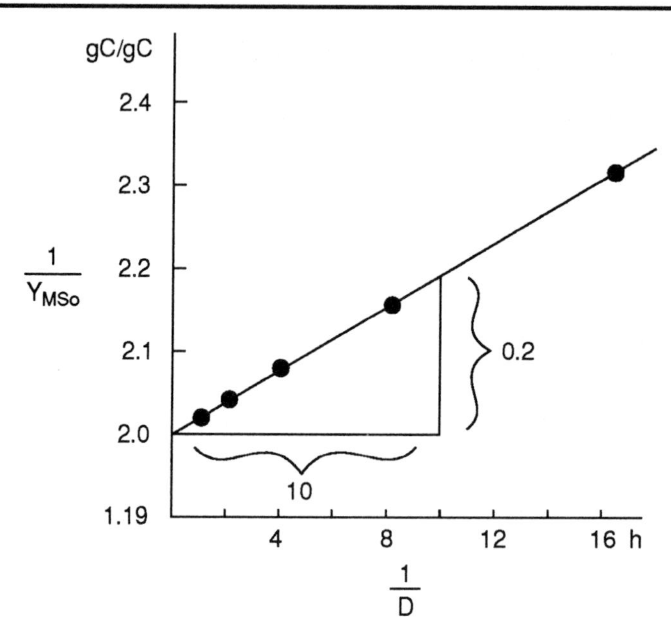

slope = r_m = $\frac{0.2}{10}$ = 0.02g of substrate C (g biomass C)$^{-1}$

Intercept = approx 2 g C/g C = $\frac{1}{Y_{MSg}}$

Thus Y_{MSg} = 0.5 g C/g C

Thus the growth yield = 0.5 g biomass C (g substrate C)$^{-1}$ and the specific rate of maintenance energy consumption is 0.02 g substrate C (g biomass C)$^{-1}$ h^{-1}.

You should note that this method to determine Y_{MSg} and r_m is not particularly accurate. This is especially true when data points cover a limited range of values. Determining the intercept value requires extrapolation and the accuracy of Y_{MSg} is limited. This is compounded by the fact that reciprocal values of measurement data are used.

4.11 We tend to underestimate the cell yield because:

- some of the cell mass actually produced from a certain mass of substrate may have been lost due to cell decay;
- some of the substrate mass supposedly used for the production of a certain cell mass may actually have been used for cell maintenance purposes.

4.12 1) 0.5 h^{-1}

We derived this from $r_m = \frac{r_d}{Y_{MSg}}$

Thus r_d = 1 x 0.5 h^{-1} = 0.5 h^{-1}

2a) 0.33g biomass C (g substrate C)$^{-1}$

2b) 0.25g biomass C (g substrate C)$^{-1}$

Here we can use either of the two following relationships:

$$\frac{1}{Y_{MSo}} = \frac{1}{Y_{MSg}} + \frac{r_m}{D} \text{ or}$$

$$\frac{1}{Y_{MSo}} = \frac{1}{Y_{MSg}} + \frac{r_d}{Y_{MSg} D}$$

Using the former, at dilution rate of 1h^{-1}

$$\frac{1}{Y_{MSo}} = \frac{1}{0.5} + \frac{1}{1} = \frac{3}{1}$$

Thus Y_{MSo} = 0.33 g biomass C (g substrate C)$^{-1}$

At a dilution rate of 0.5 h^{-1} $\frac{1}{Y_{MSo}} = \frac{1}{0.5} + \frac{1}{0.5} = 4$

Y_{MSo} = 0.25 g biomass C (g substrate C)$^{-1}$

In this question we gave a rather high value for the specific rate of maintenance energy consumption. You can see its consequences on the actual observed growth yield. What would be the observed growth yield if the dilution rate was reduced to 0.1 h^{-1}?

Then $\frac{1}{Y_{MSo}} = \frac{1}{0.5} + \frac{1}{0.1} = \frac{12}{1}$

Thus Y_{MSo} = 0.083 g biomass C/g substrate C

In other words, at this slow dilution rate most of the substrate would be used for cell maintenance and rather little for cell growth.

This is an important lesson so it is well worth remembering.

4.13

1) Organism II - it has a low affinity (high K_s) for its substrate

2) Organism I - it has a high affinity (low K_s) for its substrate and can grow well at low substrate concentrations. Low substrate concentrations are said to provide oligotrophic growth conditions.

3) Organism I - approximately 1.2 h^{-1}; organism II = 1.1 h^{-1}:

For organism I, since,

$$\mu = \frac{\mu_{max} C_{Si}}{K_s + C_{SI}} \text{ (Monod relationship)}$$

then

$$= \frac{1.2 \times 10}{0.01 + 10} = 1.2\ h^{-1}$$

For organism II, $C_{Si} = K_s$, thus the culture will grow at $\frac{1}{2}\ \mu_{max} = 1.1\ h^{-1}$.

4) Organism I - at a substrate concentration of 0.05 g C m^{-3} the specific growth rates of organism I and II will be 1 h^{-1} and 0.011 h^{-1} (calculated from the Monod relationship, see 3) above).

5) Organism II - at a substrate concentration of 20 g C m^{-3} the specific growth rate of organisms I and II will be 1.2 h^{-1} and 1.47 h^{-1} (again calculated from the Monod relationship).

6) Yes, 11.98 g sub C m^{-3}.

For the two growth rates need to be equal to allow co-existence then:

$$\frac{\mu_{max1}\ C_S}{C_S + K_{S1}} = \frac{\mu_{max2}\ C_S}{C_S + K_{S2}}$$

Substituting in the given values

$$\frac{1.2\ C_S}{C_S + 0.01} = \frac{2.2\ C_S}{C_S + 10}$$

dividing both sides by C_S gives

$$\frac{1.2}{C_S + 0.01} = \frac{2.2}{C_S + 10}$$

Therefore $2.2\ C_S + 0.022 = 1.2\ C_S + 12$

$1\ C_S = 12 - 0.022$

$= 11.998$ g substrate C m^{-3}

Responses to Chapter 5 SAQs

5.1

1) For organism A, the iodophase lasts for the period from about 10 to about 20 h after inoculation. For organism B, the iodophase lasts for the period from about 18 to about 24 h^{-1} after inoculation. Remember that iodophase represents the period during which the product is made.

2) For organism A, the trophase lasts for the period from about 8 to about 20 h after inoculation, for organism B, this phase also lasts from about 10 to about 20 h after inoculation. Remember that the trophase is the growth phase.

3) Product A appears to be linked to the growth of organism A (the product curve and growth curve appear to follow each other, except that the product is stable and does not decline along with the cell mass). Product A therefore appears to be a primary product. Product B on the other hand only starts to be made when the growth of organism B is more or less complete and, by definition, would appear to be a secondary metabolite.

4) Product B. Since the production of A by organism A is growth linked, to produce A we need to have organism A in continuous growth, this would best be achieved in a CFSTR. In contrast product B is maximally produced when the cells reach the end of the log phase. Thus to produce B we need to get the cells into this phase. This can be achieved in batch culture, whereas a CFSTR would maintain the cells in a growth phase.

You may however have been rather clever here and suggested that organism B could be grown up in a CFSTR (ie in its trophase) and transferred to a batch system to enter its iodophase. The temporal separation of the trophophase and iodophase does allow one to think about the possibilities for physically separating them into a two stage process.

5.2

	CFSTR	Batch reactor
Is it open or closed?	open	closed*
Is the environment constant or varying?	constant	constantly varying
Is it operationally complex or simple?	complex	relatively simple

* Batch reactors are usually regarded as closed as nutrients are not added during operation. However in practice they are often not strictly closed systems as gases especially oxygen and carbon dioxide may be exchanged between the reactor and the outside.

5.3

1) True - see Section 5.4 if in doubt.

2) True - of course if the culture becomes contaminated or a piece of equipment breaks, then the productivity of the CFSTR will be altered.

3) True - it is obviously growth linked and therefore, by definition, a primary product.

4) True, see Section 5.4 if in doubt.

Responses to Chapter 6 SAQs

6.1

	CFSTR	BR	PFR
Mixing	fully	fully	in axial direction, none in tangential direction
Process continuity	continuous	discontinuous	continuous
Reaction environment	constant	constant	varying in tangential direction

You may not have used exactly the same terms we have used but you should be able to judge whether or not your response was correct in each case.

6.2

No, one can only calculate the substrate consumption rate in a PFR when the substrate concentration became zero exactly when passing the point of the reactor outlet. When measuring a zero substrate concentration at the outlet, this concentration could have been reached long before the point of the outlet was reached.

6.3

1) 0.28 g biomass C m^{-2}

The relationship we use is $v_z \dfrac{dC_{Mz}}{dz} = \mu C_{Mz}$ (Equation 6.11)

Thus $v_z = 0.1$ m s^{-1} (since $v_z = \dfrac{F}{A}$)

$\mu = 1h^{-1} = \dfrac{1}{3600} s^{-1}$ (given)

$C_{MZ} = 100$g biomass Cm^{-3} (given)

Thus rate of change of biomass concentration $= \dfrac{dC_{Mz}}{dz} = \dfrac{1 \times 100}{3600 \times 0.1}$

= 0.28g biomass Cm^{-2}

2) 0.28 g product Cm^{-2}

The relationship we use is $v_z \dfrac{dC_{Pz}}{dz} = r_P C_{Mz}$ (Equation 6.12)

Thus rate of change in product concentration $\dfrac{dC_{Pz}}{dz} = \dfrac{1 \times 100}{3600 \times 0.1}$

(since r_P = 1g product C/g biomass C h = $\dfrac{1}{3600}$ g product C/g biomass s)

Thus $\dfrac{dC_{Pz}}{dz}$ = 0.28g product m^{-2}

3) The relationship we need is:

$$v_z \frac{dC_{Sz}}{dz} = -\mu \frac{C_{Mz}}{Y_{MSo}} - r_P \frac{C_{Mz}}{Y_{PS}} \quad \text{(Equation 6.10)}$$

$v_z = 0.1 \text{ m s}^{-1}$; $\mu = 1 \text{ h}^{-1} = \frac{1}{3600} \text{ s}^{-1}$; $C_{Mz} = 100\text{g m}^{-3}$

$Y_{PS} = 0.1$g product C/g substrate C, $r_P = \frac{1}{3600}$ g product C/g biomass s and

$$d\frac{C_{Sz}}{dz} = -1200\text{g m}^{-4}$$

(All of these values were given in the question)

Substituting in the above equation

$$0.1 \times -1200 = -\frac{1 \times 100}{Y_{MSo}} - \frac{1 \times 100}{3600 \times 0.1}$$

$$-120 = -\frac{100}{Y_{MSo}} - 0.27$$

$$-119.73 = -\frac{100}{Y_{MSo}}$$

Thus $\frac{1}{Y_{MSo}} = 1.1973$ C/g substrate C

$Y_{MSo} = 0.835$g biomass C/g substrate C

The key to success in carrying out these types of calculations is to make certain that each of the components in the relationships are expressed in the same units. Thus either use a time base of s or h, not a mixture of both.

Responses to Chapter 7 SAQs

7.1

0.51 g biomass C/g substrate C

Since no product, other than biomass is made, Equation 7.7 reduces to:

$$D \frac{C_{Si} - C_{Sl}}{C_{Ml}} = \frac{\mu}{Y_{MSo}}$$

$$\text{Thus } 2 \frac{(500 - 10)}{500} = \frac{1}{Y_{MSo}}$$

$$\frac{1}{Y_{MSo}} = 1.96$$

Y_{MSo} = 0.51g biomass C/g substrate C

7.2

1) 2

We can use the relationship:

$D [1 + \alpha - \alpha \beta] = \mu$ (Equation 7.8) where:

$D = 2h^{-1}$, $\alpha = 40\% = 0.4$ and $\mu = 1.2$

Thus $2 [1 + 0.4 - 0.4 \beta] = 1.2$

$2.8 - 0.8 \beta = 1.2$; $0.8 \beta = 1.6$; $\beta = 2$

2) 250g biomass C m^{-3}

Since β (concentration factor) $= \frac{C_{Mc}}{C_{Ml}} = 2$ and $C_{Mc} = 500g\ m^{-3}$

then C_{Ml} = 250 g biomass C m^{-3}.

7.3

0.8 h^{-1}

Since $D [1 + \alpha - \alpha B] = \mu$ (Equation 7.8)

then $2 (1 + 0.2 - 0.2 \times 4) = \mu$

$2(0.4) = \mu$

$\mu = 0.8\ h^{-1}$

Thus in a system fitted with cell recycling μ can be significantly smaller than D

7.4

cell mass production rate	$F_c \cdot C_{Mc}$
specific cell mass production rate	$\frac{F_c}{V} \cdot \frac{C_{Mc}}{C_{Ml}}$
sludge age	$\frac{F}{F_c} \cdot \frac{C_{Ml}}{C_{Mc}} \cdot \theta_h$
mean cell residence time	$\frac{1}{\mu}$
hydraulic residence time	$\frac{V}{F} \left(= \frac{1}{D} \right)$
concentration factor	$\frac{C_{Mc}}{C_{Ml}}$
total cell mass	$V \cdot C_{Ml}$

Responses to Chapter 8 SAQs

8.1

1) Immobilised enzymes. Enzyme purification is often costly.

2) Immobilised cells. These may contain many metabolic pathways the enzymes of which compete for the substrate.

3) Immobilised enzymes. With these systems we get a single (or limited number of) reaction(s), thus the product is purer. Also immobilised cells tend to make or release a variety of by-products.

4) Immobilised cells. Whole cells have functional metabolic pathways and can carry out sequential modification of substrates to products. For example immobilised yeast might transform glucose to ethanol. Using separate enzymes makes this practically impossible.

5) Immobilised cells. Cells contain many different enzymes. The enzyme we might wish to use might only be present in minute quantities within the biomass.

6) Immobilised cells. With these systems, the substrate we wish to use will first have to penetrate the cell membrane. Thus although cells actively take up some substrates (normally the substrates they grow on) some compounds will not penetrate the membrane. Say for example we wish to transform glyceraldehyde-3-phosphate to glyceric acid-3-phosphate using a biocatalyst, these phosphorylated intermediates will not penetrate the cell membrane. We would either have to use purified enzymes or make the cell membranes more permeable (by, for example, using a detergent). Making the cell membrane more permeable almost inevitably means that the cell dies.

8.2

Your selection of definitions should have been:

Words	Definitions
attachment	cells sticking to other cells
sloughing	detachment of clumps of cells from a biofilm
substratum	a surface onto which cells stick
motility	cells possessing a mechanism of locomotion
taxis	cells move up a gradient of an attracting chemical (postive chemotaxis)
detachment	cells loosening from a biofilm
desorption	cells loosening from a substratum
adsorption	cell sticking to uncolonised substratum
erosion	detachment of single cells from a biofilm

8.3

1) increase

2) increase

3) decrease

4) decrease

We can come to the right answers by using Equation 8.5b. However instinctively we should also come to the correct answer to this question.

Thus we would anticipate the faster the cells are growing, the greater the consumption of substrate. Thus we would expect higher growth rate to create a steeper gradient. Likewise, the more biomass present, the greater the rate of substrate consumption and thus the steeper the gradient. A higher grow yield means less substrate is used to produce a fixed quantity of biomass, and this will result in a less steep gradient. Likewise a higher rate of diffusion means the substrate will spread through the film rapidly. This will also tend to reduce the gradient.

We may conclude that equations of the type given in Equation 8.5b are consistent with what common sense would tell us.

8.4

1) $0.5\ h^{-1}$

The relationship we need is:

$$\frac{dC_{Mf}}{dt} = -\ C_{Mf} \cdot r_{Md} + C_{Mf} \cdot \mu_f \text{ (Equation 8.10)}$$

But at a steady state the net rate of cell mass accumulation $\left(\frac{dC_{Mf}}{dt}\right)$ is zero.

Thus $C_{Mf}\ r_{Md} = C_{Mf}\ \mu_f$

In other words the net rate of cell mass detachment = rate of biofilm cellular growth

or $r_{Md} = \mu_f$

Thus the specific cell detachment rate = $0.5\ h^{-1}$

From the result given above we can see that in a steady state $r_{md} = \mu_f$. We do not therefore need to know the biomass concentration in the film providing we know the value of μ_f.

2) $0.1\ h^{-1}$

The relationship we need is:

$$\frac{dC_{Pl}}{dt} = D\,(C_{Pi} - C_{Pl}) + C_{Pf} \cdot \frac{A}{V}\, r_{Pd} + C_{Ml} \cdot r_{Pl} \text{ (Equation 8.8)}$$

This is just the mathematical form of:

net rate of product accumulation	=	net rate of product inflow	+	rate of product detachment	+	rate of product formation in liquid phase

But in a steady state $\frac{dC_{Pl}}{dt} = 0$

Also, since there is no biomass in the liquid $C_{Ml} \cdot r_{Pl} = 0$.

Thus $0 = D(C_{Pi} - C_{Pl}) + C_{Pf} \frac{A}{V} r_{Pd}$

$0 = 0.2\ (0 - 400) + 800 \cdot \frac{1}{1} r_{Pd}$

$80 = 800\ r_{Pd}$

Thus $r_{Pd} = 0.1\ h^{-1}$

3) $1000\ g\ C\ m^{-3}$

For the liquid:

the rate of substrate accumulation	=	net rate of substrate inflow	-	rate of substrate uptake by biofilm	-	rate of substrate uptake in liquid phase for cells and products

(see Equation 8.6)

At steady state:

$0 = D(C_{Si} - C_{Sl}) - 2000g\ C\ m^{-2}\ h^{-1} - 1000g\ C\ m^{-3}\ h^{-1}$

(given) (given)

$0 = 1\ (4000 - C_{Sl}) - 2000 - 1000$

Thus $C_{Sl} = 1000g$ substrate $C\ m^{-3}$

8.5

1) 0.092 g product C/g substrate C

Since all of the product is growth linked then $k_{Pn} = 0$ and $r_P = k_{Pg}\,\mu$

Thus $k_{Pg} = \frac{r_P}{\mu} = \frac{0.2}{0.5} = 0.4$

But $r_{Sf} = \mu \left[\frac{1}{Y_{MSo}} + \frac{k_{Pg}}{Y_{PS}} \right] + \frac{k_{Pn}}{Y_{PS}}$ (Equation 8.15)

Thus:

$$3.834 = 0.5\left[\frac{1}{0.3} + \frac{0.4}{Y_{PS}}\right] + 0$$

$$3.834 - 1.667 = \frac{0.2}{Y_{PS}}$$

$$Y_{PS} = \frac{0.2}{3.834 - 1.667} = 0.092 \ \frac{\text{g product C}}{\text{g substrate C}}$$

2) No we cannot conclude that biomass and product are the sole derivatives of the substrate. Since $Y_{MSo} = 0.3$ g biomass C/g substrate C and $Y_{PS} = 0.092$ g product C/g substrate C it means we can only account for 0.392 (ie 0.3 + 0.092) g of C for every g of substrate C. But C cannot be created or destroyed, so some of the carbon from the substrate has not been accounted for. In all probability this is released from the system as CO_2 and not taken into account even as a product.

Responses to Chapter 9 SAQs

9.1

1) 10 minutes (approximately). It is read from the graph by determining how long the delay is between when the pump should have been switched on and when it actually was. Thus on the graph:

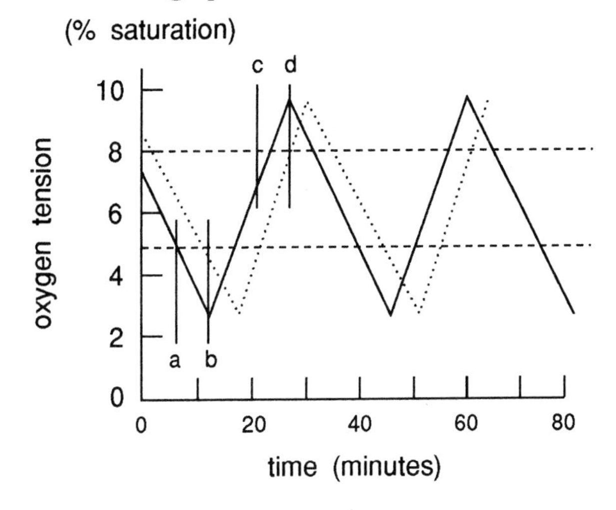

a) pump should be switched on here

b) pump actually switched on here

c) pump should be switched off here

d) pump actually switched off here

2) We would expect a massive overshoot of oxygen in the bioreactor since the input rate of oxygen would increase (approximately 5 fold) when pure oxygen replaced air.

Graphically this would be:

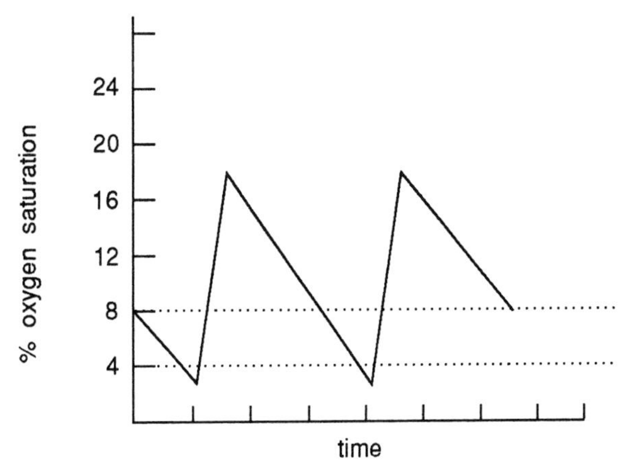

3) The best way to achieve the desired result would be to reduce the delay time in the control circuit. If this could be reduced to, for example, a few seconds then there would be no overshoot. There are other ways to achieve the desired results (eg narrowing the range of the set points) but this is unreliable and not recommended).

9.2

1)

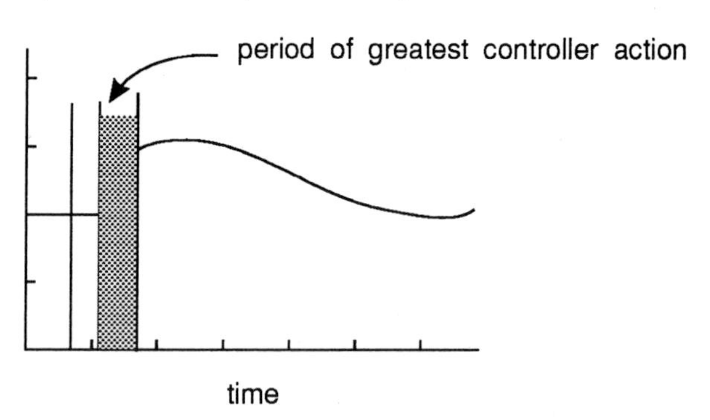

Essentially what the differentiating element detects is the rate of change of the measured output.

ie $\frac{d[O]}{dt}$ where [O] = output signal and t = time

Graphically the data above would produce:

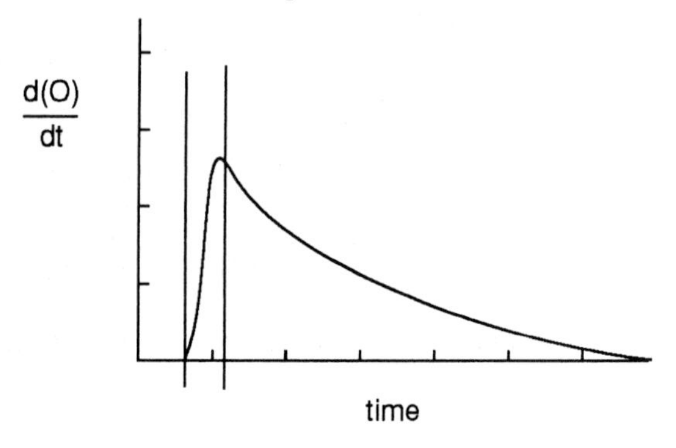

9.3

1) True (see Section 9.4.3)

2) False. Integrating elements need time in order for the signal to build up. They are, therefore usually slow and have a tendency to cause an overshoot in the response.

3) True (see Sections 9.2.1 and 9.4.1).

4) False. Although a long time delay between changes in measured values and control action can cause overshoot - see 3) - a very short delay is not always desirable. With very short delays, the controller may to become very sensitive to noise in the measured value. This sensitivity is described as making the controller nervous.

5) Usually false. In principle proportional controllers can maintain the output of a process at a set point without re-adjustment. In practice, however we may experience drift (see Section 9.4.1).

6) True.

7) True. We are inferring the value of a variable from the measured value of another variable.

9.4 The variables that will need controlling are:

- temperature - since parameters a, b and m are temperature sensitive.
- substrate concentration - since parameters a and b are sensitive to substrate concentration.
- specific growth rate - since the relationship between product formation and substrate utilization depends upon the specific growth rate (μ).

Temperature could be controlled by a typical thermostat. Since substrate cannot be measured directly on-line, we would have to use an inferential method. An obvious way is to measure the product in the process output and to apply a state estimator using the relationship given in the question to estimate the substrate concentration. This estimated substrate concentration could be used to control the substrate feed. In order for the state estimator to calculate the substrate concentration, it will need values of a, b, m and μ. The determination of μ is itself difficult on-line. It can be determined from the measurement of biomass concentrations. We will be examining on-line biomass determination in the next section so we will not go into further details here except to say that μ can be determined from measurement of biomass. Note however that μ is sensitive to substrate concentrations and temperature. Thus in order for the state estimator to do its job effectively, it needs to have information on all three variables (temperature, μ and substrate concentration). The first of these it may get directly from a temperature sensor, the second indirectly (inferential) by measuring biomass and the third by measuring product. We would also need to know how parameters a, b μ and m are influenced by temperature and how parameters a and b are sensitive to substrate concentrations. Without this knowledge we would still not be able to predict q_p and q_s.

Thus the sort of control network we might envisage being set up can be illustrated as follows:

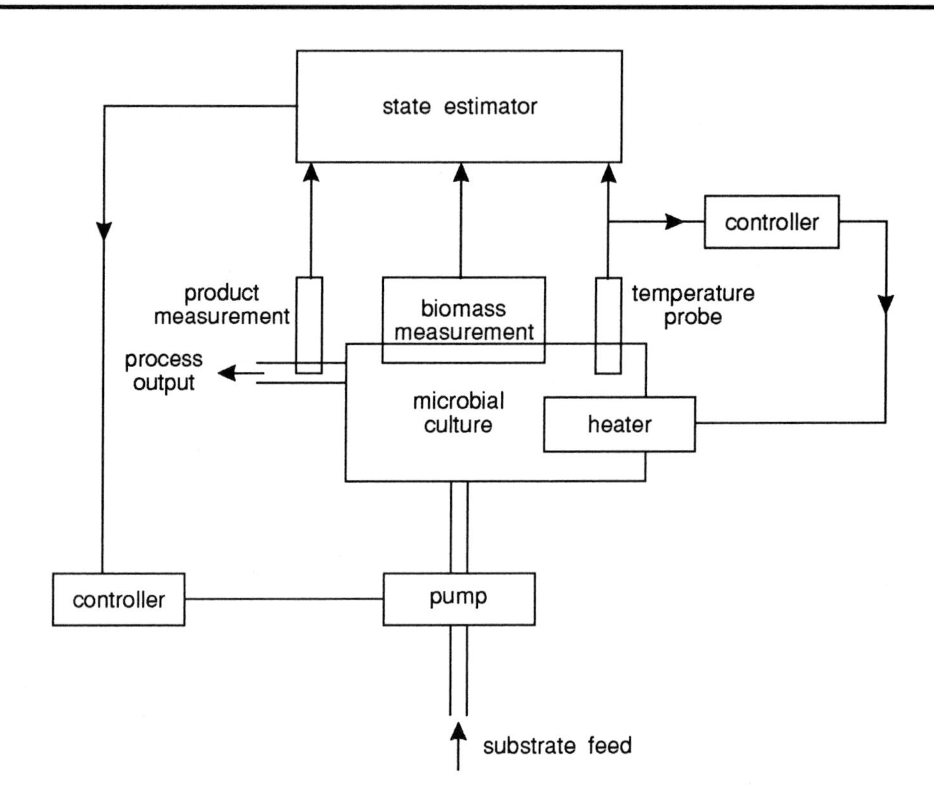

There is some scope as to the exact arrangement and to the exact types of measurements that are used. For example the state estimator may not necessarily be directly coupled to the temperature probe. If it was not so connected it would need to be supplied with values of a, b and μ. If however it was provided with an on-line measurement of temperature, it could use this to give a continual estimation of a, b, and μ.

9.5

1) unstable;

2) specific;

3) column fouling;

4) non-linear.

Suggestions for further reading

Chapter 2

Kossen N W D and Oosterhuis M N G (1985)
'Modelling and scale up of bioreactors' in biotechnology vol 2 Ed Rehm H J & Reed G, BCH Verlagsgesellschaft, Weinheim

Chapters 4 - 8

BIOTOL text 'Operational Modes of Bioreactors' (1992)
Butterworth-Heinemann, Oxford

Birnbaum S, Larsson P O and Mosbach K (1986)
'Immobilised biocatalyst, the choice between enzymes and cells' in Webb C, Black G M and Atkinson B 'Process engineering aspects of immobilised cell systems'. The institution of chemical engineers, 35-53

Fera P, Siebel M A, Characklis W G and Prieur D (1989)
'Seasonal variations in bacterial colonisation of stainless steel, aluminum and polycarbonate surfaces in a seawater flow system'. Biofouling, 1 . 251-261

Jang L K and Yen T F (1985)
'A theoretical model of convective diffusion of motile and non-motile bacteria toward solid surfaces' in Zajic J E and Dpma;dspm E C (Eds), Microbes and oil recovery, vol 1, int'l Biosources Journal, 1985, 226-246

Luedeking R and Piret E L (1959)
'A kinetic study of the lactic acid fermentation'. Batch process at controlled pH, J of Biochem and Microb Techn and Engrg, 1 (4), 393-412

Monod J (1949)
'The growth of bacterial cultures'. Annual Rev of Microbiol, 3, 371-394

Moser A (1988)
'Bioprocess Technology', Springe Verlag, New York

van't Riet, K and Tramper J (1991)
'Basic Bioreactor Design', Marcell Dekker, New York

Roels J A and Kossen N W F (1978)
in Bull M J (Ed), 'Progress in industrial microbiology', vol 14 p 95, Elsevier, Amsterdam

Trulear, M G (1983)
'Cellular reproduction and extracellular polymer formation in the development of biofilms', PhD thesis, Montana State University, Bozeman, MT, USA

Chapter 9

Kwakernaak H and Sivan R (1972)
'Linear optimal control systems', Wiley, London
(remarks: book discusses time domain approach towards control engineering)

Dorf R C (1980)
'Modern control systems', Addison-Wesley Publishing Company, Amsterdam
(remarks: book mainly concerns s-domain approach towards control engineering)

Gelb A (1982)
'Applied optimal estimation', The M. I. T. Press, London
(remarks: book concerns state estimation methods in time domain, eg Kalman)

Halme A (1983)
'Modelling and control of biotechnological processes', Oxford Press
(remarks: book especially concerns the biotechnological field)

Appendix 1

Application of the Gauss-Jordan reduction method to dimensional analysis

In the main text (Chapter 2), we described an example of the application of dimensional analysis to scale up. In that example, the Rayleigh method was used. In this appendix, we describe a second example.

This second example uses a similar approach as the first one except that the Gauss-Jordan reduction method is used to obtain the dimensionless groups. We will use a practical example to introduce this approach. The performance of a large waste water purification plant appeared to be less than expected. Flow problems (short circuiting) were assumed to be the cause. Residence time distribution (RTD) measurements on the prototype scale confirmed this assumption. To solve the problem a model was built with a linear scale ratio of 1/20. Figure 1 schematically shows a relevant part of the system.

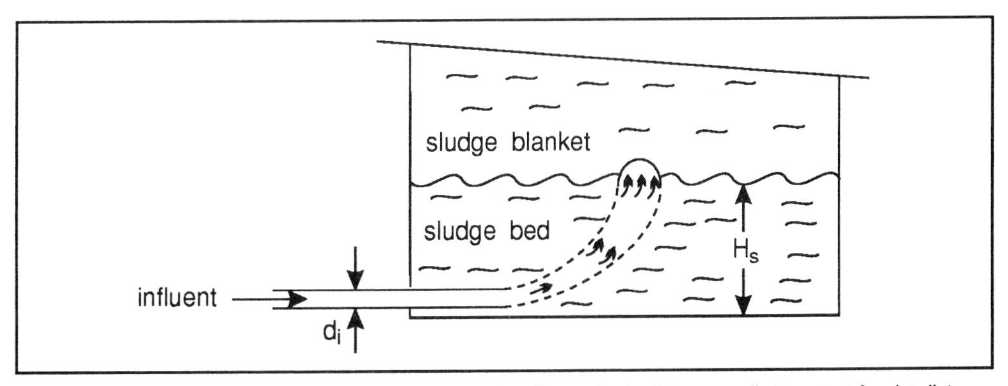

Figure 1 Schematic representation of the system under discussion in this appendix (see text for details).

The problem could be localised as short circuiting of the sludge bed.

Π Even with this brief description see if you can identify the major parameters involved in the circulation of material in a sludge bed. (Hint, it may be useful to think in terms of the parameters relating to the liquid, the bed which contains solid particles, the gas and the reactor.)

We would not anticipate that you would have identified all of the parameters involved because our description of the system was limited. Nevertheless you should have identified many of those listed below.

In practice the parameters which were selected were:

- liquid (density ρ_l, viscosity η, the velocity of the liquid v)
- bed (the particle diameter d_p, the density of the particle ρ_p, the yield stress τ_o)
- gas (superficial gas velocity v_s)
- reactor (the height of the sludge bed H_s, the internal diameter d_i of the inlet)
- general (gravitational constant g, the time t)

The dimensionless groups will be obtained by writing the variables and the dimensions in one matrix. The left part of this matrix has to be transformed into a diagonal matrix. The groups are then easily obtained, as will be shown.

	ρ	H_s	t	g	d_i	v_s	d_p	$\Delta\rho$	τ_o	v-	η
M	1	0	0	0	0	0	0	1	1	0.	1
L	-3	1	0	1	1	1	1	-3	-1	1	-1
t	0	0	1	-2	0	-1	0	0	-2	-1	-1

where $\Delta\rho$ is the difference between ρ_p and ρ_l . Addition of three times the first row to the second row results in:

	ρ	H_s	t	g	d_i	v_s	d_p	$\Delta\rho$	τ_o	v	η
M	1	0	0	0	0	0	0	1	1	0	1
L	0	1	0	1	1	1	1	0	2	1	2
t	0	0	1	-2	0	-1	0	0	-2	-1	-1

The dimensionless groups are now obtained as follows: g is divided by H_s and multiplied by t^2. Mass does not appear in this group because at the place of the column of g that corresponds within the left matrix, a zero is found. The first dimensionless group now is $\frac{gt^2}{H_s}$

In the same way the other groups are found:

$$\pi_1 \equiv \frac{gt^2}{H_s} \quad \pi_2 \equiv \frac{d_i}{H_s} \quad \pi_3 \equiv \frac{v_s\,t}{H_s} \quad \pi_4 \equiv \frac{d_p}{H_s}$$

$$\pi_5 \equiv \frac{\Delta\rho_p}{\rho} \quad \pi_6 \equiv \frac{\tau_o t^2}{\rho\,H_s^2} \quad \pi_7 \equiv \frac{v\,t}{H_s} \quad \pi_8 \equiv \frac{\eta\,t}{\rho\,H_s^2}$$

For the sake of better physical interpretation the following rearrangements are made:

π_7 , π_8 and π_2 are recombined into $\left(\pi_7 \cdot \frac{1}{\pi_8} \cdot \pi_2\right)$

$$\pi_8' \equiv \frac{\rho\,v\,d_i}{\eta}$$

π_1, π_5 and π_7 are recombined into $\left(\pi_5 \cdot (\pi_7)^2 : \pi_1\right)$

$$\pi_1' \equiv \frac{\rho v^2}{\Delta \rho g H_s}$$

Likewise π_6 and π_7 can be recombined into $\left(\pi_6 \cdot (\pi_7)^{-2}\right)$

$$\pi_6' = \frac{\tau_o}{\rho v^2}$$

Table 1 shows the final set of dimensionless numbers, together with their values at prototype and model scale.

	P	M
$\pi_1' = \frac{\rho v^2}{\Delta \rho g H_s}$	2.88	2.88
$\pi_2 \equiv \frac{d_i}{H_s}$	$1.9 \cdot 10^{-2}$	$1.9 \cdot 10^{-2}$
$\pi_3 \equiv \frac{v_s t}{H_s}$	$9.1 \cdot 10^{-2}$	-
$\pi_4 \equiv \frac{d_p}{H_s}$	10^{-3}	10^{-3}
$\pi_5 \equiv \frac{\Delta \rho}{\rho}$	10^{-1}	10^{-1}
$\pi_6' \equiv \frac{\tau_o}{\rho v^2}$	$2.9 \cdot 10^{-3}$	-
$\pi_7 \equiv \frac{v t}{H_s}$	$1.1 \cdot 10^{-3}$	-
$\pi_8' \equiv \frac{\rho v d_i}{\eta}$	$9.1 \cdot 10^{4}$	1029

Table 1 Dimensionless Numbers derived from the problem. (Adopted from Kossen and Oosterhuis,1985, 'Biotechnology' 2p 572. Ed. Rehmand, Reed, VCH Verlagsgesellschaft, Weinheim.

As one was not especially interested in the effect of time, the groups π_3 and π_7 were neglected. The group π_6 is so small that it is very unlikely that this group is of any importance (the turbulent shear stress is much larger than the yield stress of the bed); group π_6 was therefore also neglected. π_2 and π_4 were kept constant (geometric similarity). The most important group was π_1. The physical interpretation of this group has been mentioned in a previous section. As a consequence of keeping π_1 constant, the value of the Re number decreased considerably (π_8). Because it was clearly observed that despite this low Re number the jet at model scale was still turbulent, this low Re number was no limitation for the procedure.

Improvements of the flow system were performed in the model for π_1 = constant. The solutions proved to be successful when translated backwards to the prototype.

Appendix 2

A mathematical approach to systems dynamics, process control and optimisation

In the main body of the text of Chapter 9 (Section 9.3.2), we explained the importance of being able to linearise systems in the analysis of systems dynamics. In this appendix we focus on the important tools to analyse continuous, linear, time independent, single input systems. Although this may seem somewhat restricted, many of these tools can be applied to many other processes.

Important in this approach is the Laplace transformation. This is a widely used tool used in linear systems analysis. In summary, the Laplace transformation transforms a function in the time domain into the Laplace domain (often termed the s-domain or the frequency domain). A major advantage of this is that we can work with simple algebraic equations (in s) instead of differential equations (in t). Due to the properties of the Laplace transformation, integration, differentiation and coupling of different parts of the system become quite easy.

In the next section, we have provided some guidelines on the Laplace transformation. These guidelines are based on the discussion of P. G. Doucet and P. B. Sloep (1992 'Mathematical Modelling in the Life Sciences' Ellis Horwood, Chichester).

If you are already familiar with this mathematical tool, you may find that the guidelines provided will serve as a quick form of revision.

1 The Laplace transformation

Linear systems are important in control engineering as they often allow for mathematical analysis of the behaviour of the system. In solving the set of differential equations that describes the system Laplace transformations constitute a powerful tool.

The Laplace transformation, which we will denote by *L*, maps an entire function into another function. The reader should note that a variety of symbols are used in the literature to indicate Laplace transformations, usually based on the letter L. By this transformation both sides of the differential equation are transformed into algebraic equations that can be solved by ordinary algebra. Once solved the algebraic equation is transformed back, using a table of transforms. Powerful as this tool may be, there is one crucial restriction to its use: it only works with systems of linear differential equations. In working with linear control systems this poses of course no problems.

The recipe for the Laplace transformation is as follows. Multiply F, which is a function of time, by an exponential function of two variables, e^{-st}, and subsequently integrate the product from $t = 0$ to ∞. The function obtained by this transformation is denoted f and is called the Laplace transform of F:

$L(F) = f$

with:

$$f: s \rightarrow \int_{t=0}^{t=\infty} F(t)\, e^{-st}\, dt$$

As the exponent of e^{-st} must be dimensionless, the variable s has the dimension $time^{-1}$.

It will be evident that f depends on s and on the function F, but not on t. From the definition of *L* it follows that *L* maps an entire function into another function, and therefore any particular f-value is determined by the entire function F. When we think for a moment about this, it will be clear that if we know only a portion of F, we do not know f and, the other way around, in order to know one point of F we must know the complete f.

Another point to mention is that in theory not all functions F possess a transform. However, for most practical applications the functions encountered will possess a Laplace transform. Finally we state without proof that *L* is one-to-one mapping which means that if F has the transform f, then there is no other function which also has f for its transform, and hence the inverse transform L^{-1} exists.

Let us now illustrate the principle of the Laplace transformation by applying it to a simple function. Let us consider the function $F(t) = e^{-kt}$

Applying the definition of *L* yields:

$$f(s) = \int_{0}^{\infty} e^{-kt}\, e^{-st}\, dt = \int_{0}^{\infty} e^{-(s+k)t}\, dt$$

which gives:

$$f(s) = \frac{-1}{s + k}$$

It will not always be easy to obtain a transform of a particular function by applying the definition of *L* given above, but there are rules allowing us to find it.

Sums and scalar products

When $L(F) = f$ and $L(G) = g$ then $L(F + G) = fg + g$

Furthermore $L(\alpha F) = \alpha f$

Thus by combination $L(\alpha F + \beta G) = \alpha f + \beta g$ which means that L is a linear operator.

Derivatives

$L[F'(t)] = sf(s) - F(0)$

repeated use of this rule enables us to find the transform for the second (and higher order) derivative:

$$L(F'') = s^2 f(s) - sF(0) - F'(0)$$

Integral

$$L\left(\int_0^t F(t)\,dt\right) = 1/s\ f(s)$$

Time shift

When a function G(t) is obtained from function F(t) by a time shift a (thus G(t) = F(t-a) - see Figure 1) then:

$$L(G) = g(s) = e^{-as} f(s)$$

on the condition that the shifted function equals 0 over the first a time units.

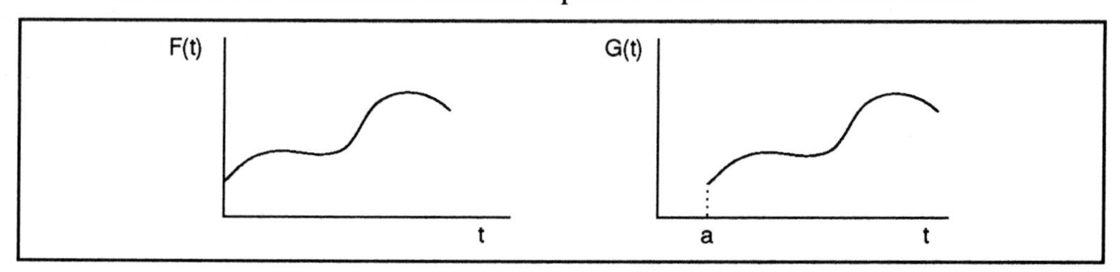

Figure 1 Time shift: G(t) = F(t - a). For t between 0 and a, G(t) is defined as 0.

Convolution

We have seen above that sums and multiples of functions carry over to the transforms, but products do not, thus F(t) . G(t) does not have the transform f(s) . g(s). The combination of F(t) and G(t) which does have the transform f(s) . g(s) is called the convolution of F and G:

$$L\left(\int_0^t F(t-\tau)\,G(\tau)\,d\tau\right) = f(s)\,.\,g(s)$$

Initial value theorem

$$\lim_{t \to 0} F(t) = \lim_{s \to \infty} sf(s)$$

Final value theorem

$$\lim_{t \to \infty} F(t) = \lim_{s \to 0} sf(s)$$

Some examples of Laplace transformation

Let us consider the simple differential equation describing the rate of substrate conversion in a process obeying first order kinetics:

$C'(t) = - k\,C(t)$ with the initial condition $C(0) = C_0$

The left-hand and right-hand sides both denote functions, and if C(t) is a solution they both denote the same function, implying that their Laplace transforms must be equal. Now let *L* C(t) be c(s) then applying the rule for derivatives to the left-hand side and the rule for sums and scalar products to the right-hand side will give us:

$$sc(s) - C_0 = - kc(s)$$

which can be rearranged to:

$$c(s) = \frac{C_0}{s + k}$$

This means that we have found the transform of C and now we only have to find the inverse transform of c. In the simple example given above we found that $1/(s + k)$ is the transform of e^{-kt}, and thus we arrive at:

$$C(t) = C_0\, e^{-kt}$$

which is the well know expression of a first order process.

We can summarise the procedure for Laplace transformation as follows:

Step 1 - translation of the differential equation into an algebraic equation

Step 2 - solving of the algebraic equation resulting in an expression for c

Step 3 - back transformation of c to the solution C

When the procedure is applied to a set of differential equations the procedure remains essentially the same: the set of differential equations is transformed into a set of algebraic equations, these equations are solved and the solutions are transformed back. Let us work through an example. We start with the following set of differential equations:

$Q_1'(t) = -k_1Q_1(t)$ $\qquad Q_1(0) = Q_0$

$Q_2'(t) = k_1Q_1(t) - k_2Q_2(t)$ $\qquad Q_2(0) = 0$

Transform:

$$s\, q_1(s) - Q_0 = - k_1\, q_1(s)$$

$$s\, q_2(s) - 0 = k_1\, q_1(s) - k_2\, q_2(s)$$

Solve:

$$q_1(s) = \frac{Q_o}{s + k_1}$$

$$q_2(s) = \frac{k_1 Q_0}{(s + k_1)(s + k_2)}$$

The final step will be to find the back transformations. For the first equation we have met the back transformation already, but the second equation needs to be worked on first. It can be shown that:

$$\frac{1}{(s + k_1)(s + k_2)} = \frac{1/(k_2 - k_1)}{s + k_1} - \frac{1/(k_2 - k_1)}{s + k_2}$$

Hence:

$$q_2(s) = \frac{k_1 Q_0}{k_2 - k_1}\left[\frac{1}{s + k_1} - \frac{1}{s + k_2}\right]$$

We thus have rewritten $q_2(s)$ as the sum of two familiar transforms, and therefore we can now transform it back. We obtain as a solution:

$$Q_1(t) = Q_0\, e^{-k_1 t}$$

$$Q_2(t) = \frac{k_1 Q_0}{k_2 - k_1}\left[e^{-k_1 t} - e^{-k_2 t}\right]$$

A similar procedure can be followed for a set of more than two simultaneous differential equations. If a set of simultaneous differential equations is transformed, a set of algebraic equations is obtained:

$sx_i(s)$ = a weighted sum of $x_1(s)$, $x_2(s)$ etc

Solving these equations results in a set of rational functions of s:

$$x_i(s) = \frac{\text{a polynomial of degree less than n}}{\text{a polynomial of degree n}} = \frac{\phi_i(s)}{\psi(s)}$$

where n is the number of equations. The numerator polynomal $\phi_i(s)$ is different for each $x_i(s)$, but the denominator polynomal $\psi(s)$ is the same for all and is known as the system's characteristic polynomal.

Once a solution is found, it must be transformed back. In the previous example the denominator polynomal was of degree 2 and could be written as the sum of partial fractions; the same strategy is followed in the present example. The denominator polynomal is decomposed to:

$$\psi(s) = (s - \lambda_1)(s - \lambda_2) \ldots (s - \lambda_n)$$

in which the λ's are the eigenvalues of the system's coefficient matrix. The solution can now be written as:

$$x_i(s) = \frac{A_1}{s - \lambda_1} + \frac{A_2}{s - \lambda_2} + \ldots + \frac{A}{s - \lambda_n}$$

and back transformation yields:

$$X_i(t) = A_1 e^{-\lambda_1 t} + A_2 e^{-\lambda_2 t} + \ldots + A_n^{e-\lambda_n t}$$

The following points must be mentioned:

- the functions $X_1(t),\ldots,X_n(t)$ each have their own set of coefficients $A_1,\ldots,A_n$, but all share of course the same set of exponents $\lambda_1, \ldots, \lambda_n$

- some of the λ's may be complex numbers and complex exponentials result in sines and cosines;

- if some of the λ's are equal, some of the A's are no longer constant but are polynomals in t.

For most practical applications a fairly small number of transforms will suffice to solve the problems and commonly use is made of tables in which transforms are summarised. Therefore, in practice the use of Laplace transforms often assumes the form of a 'cookbook-like' procedure: transform the function by using a table, solve the algebraic equation and transform back again by using a table. Table 1 lists the most useful transforms.

We finally should mention that the Laplace transformation can not be applied to all differential equations. It can be applied to any set of linear differential equations with constant coefficients, but it does not work with non-linear equations or linear equations with time dependent coefficients.

F(t)	f(s)	F(t)	f(s)
1	$\frac{1}{s}$	$\frac{1}{a}(1-e^{-at})$	$\frac{1}{s(s+a)}$
t	$\frac{1}{s^2}$	$\frac{e^{-at}-e^{-bt}}{b-a}$	$\frac{1}{(s+a)(s+b)}$
t^n	$\frac{n!}{s^{n+1}}$	$\frac{ae^{-at}-be^{-bt}}{a-b}$	$\frac{s}{(s+a)(s+b)}$
e^{-at}	$\frac{1}{s+a}$	sin at	$\frac{a}{s^2+a^2}$
te^{-at}	$\frac{1}{(s+a)^2}$	cos at	$\frac{s}{s^2+a^2}$
$\frac{1}{(n-1)!}t^{n-1}e^{-at}$	$\frac{1}{(s+a)^n}$		
$t^n e^{-at}$	$\frac{n!}{(s+a)^{n+1}}$		
sinh (ωt)	$\frac{\omega}{s^2-\omega^2}$		
cosh (ωt)	$\frac{s}{s^2-\omega^2}$		
e^{-at} sin (ωt)	$\frac{\omega}{(s+a)^2+\omega^2}$		
e^{-at} cos (ωt)	$\frac{(s+a)}{(s+a)^2+\omega^2}$		

Unit impulse $\delta(t_0)$	1	area = 1; $t = t_0$
Unit pulse $\delta_A(t)$	$\frac{1}{A}\,\frac{1-e^{-sA}}{s}$	1/A; A

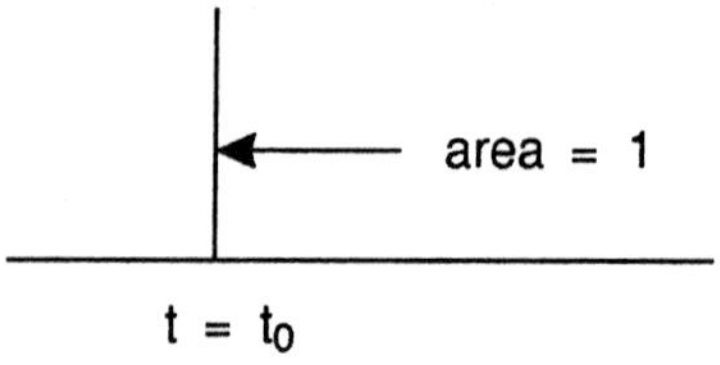

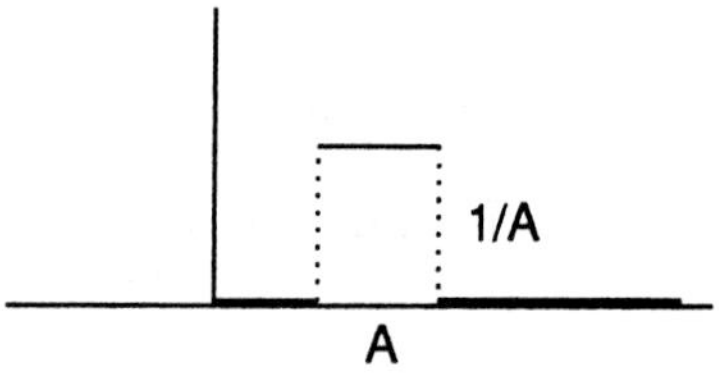

Table 1 Functions and their Laplace transforms.

1.2 Application of Laplace transformation to first order differential equations

Let us now consider the following first-order differential equation:

dx/dt = a . x + b . u

where: x: state variable, u: input variable

We are interested in the Laplace transform of x(t): L [x(t)]

According to the linearity of the Laplace transformation, the following must hold:

L [dx/dt] = a . L[x] + b.L[u]

and, according to the derivative rule:

s . L [x] - x(0) = a.L [x] + b . L[u]

Assuming that x(0) = 0, this can be rewritten to:

L[x]/L[u] = b/(s-a)

transfer function

This shows that the effect of any input signal (u) is given by the function b/(s-a) in the Laplace domain. This function is called the transfer function of the process. The transfer function is defined as the Laplace transform of the input signal(u). Usually the transfer function is denoted G(s), which stresses the fact that it is a function in s. From this transfer function, we can easily obtain L[x] when L[u] is given.

By means of an inverse transformation, the original output signal in the time domain (x(t)) can be calculated. If we assume that the input signal is a unit step (then: $L[u] = \frac{1}{s}$), we immediately know L[x]:

L[x] = (1/s) . (b/(s-a)) = b/(s . (s-a))

This can be rewritten to:

L[x] = (b/a)/(s-a) - (b/a)/s

Because the inverse Laplace transforms of these two terms are known, the latter description can simply be re-transformed to the time domain, yielding:

x(t) = (b/a) . exp(at) - (b/a)

Linear systems are generally characterised by their transfer functions. The order of the differential equation equals the order of the denominator polynomial of the transfer function. The roots of this polynomial are the poles of the system. The roots of the numerator polynomial are called zeros.

As mentioned above, the Laplace domain is also referred to as the frequency domain. To come from the s-domain in the frequency domain, one simply substitutes the variable s by (j. Ω), with $\Omega = \sqrt{(-1)}$ and Ω = frequency. Calculations now give the phase and the amplitude effect for each frequency.

2 First, second and higher order dynamics

First-order systems combine a capacity to store material or energy (or the like) with a certain resistance associated with mass or energy flow. Therefore, first-order processes possess some inertness. A good example of a first-order process is a CSTR (continuous stirred tank reactor). Assuming no reaction takes place, the differential equation for the concentration of component X is given by:

$$dC_X/dt = (F/V) . C_{Xin} - (F/V) . C_X$$

Here, the influent concentration of component X is the input signal, while C_X is considered output signal. We can now determine the transfer function of this simple process:

$$G(s) = L [C_X]/L[C_{Xin}] = (F/V / (s+(F/V))$$

Assuming that a unit step is applied on the system input, the output signal in the time domain is found to be:

$$C_X(t) = 1 - exp(-(F/V)t)$$

This clearly demonstrates the inertness of the first-order system: a sudden change on the system input is not immediately evident on the system output.

Using Laplace transformation, it can easily be shown that a second-order differential equation will yield a transfer function of the form:

$$G(s) = b/((s-a_1) . (s-a_2))$$

Π Verify the above result.

Your calculation should have been

$$d^2y/dt^2 + a . dy/dt + b . y + c = \alpha.x \rightarrow$$

$$s_2 . L[y] + a . s . L[y] + b . L[y] = \alpha . L[x] \rightarrow$$

$$L[y]/L[x] = \alpha/(s^2 + a . s + b)$$

which can be rewritten in the required form (although a_1 and a_2 may be conjugate complex)

the order of a linear system

As can be seen, the denominator polynomial has an order which equals the order of the differential equation. This is simply called the order of a linear system.

If a CSTR is connected to another CSTR, the output signal of the first is the input signal of the second CSTR. We already determined the transfer function of a single CSTR. To obtain the Laplace transform of the output signal for the given situation(step signal, two

CSTR's in series), we use the transfer function of one CSTR and multiply this by the input, which is the output of the first CSTR:

$L[C_{X2}] = L[C_{X1}] \cdot (F/V_2)/(s + (F/V_2))$

$= (1/s) \cdot (F/V_1)/(s+(F/V_1)) \cdot (F/V_2)/(s + (F/V_2))$

Note that this demonstrates the important convolution property of the Laplace transform. After some manipulations (including a partial fraction expansion), this can be re-transformed to the time domain, yielding:

$C_{X2}(t) = 1 - (V1/(V1-V2)) \cdot \exp((-F/V1)t) + (V2/(V1-V2)) \cdot \exp((-F/V2)t)$

Thus a second-order system is formed as the result of combining two first-order systems. The time domain solution now shows two exponential terms.

For a good understanding of second-order dynamics we must briefly discuss the inherently second-order dynamics in which the second order process cannot be split up into two first-order processes. A classical example of such a system is the spring-mass-damper system. Due to the inertia of motion, an overshoot can occur (see Figure 2), depending on the damping ratio of system. This type of behaviour can be recognised from the transfer function. the denominator polynomial then has two conjugate complex roots, while the time domain solution has sine and cosine functions. Due to the fact that the complex roots of the denominator polynomial are conjugate complex, the time domain solution will, again, be completely real.

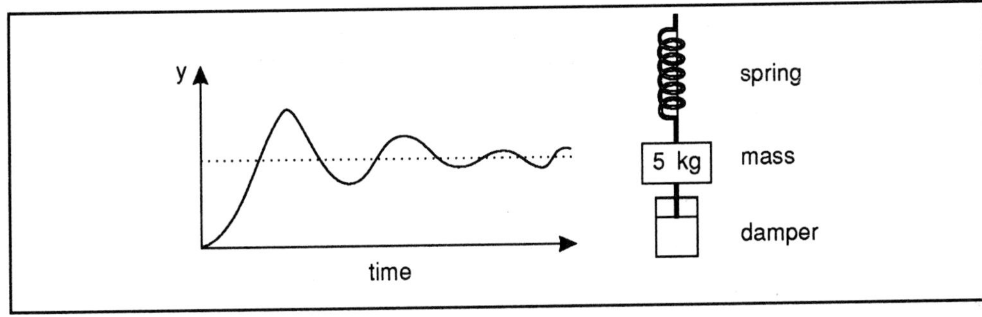

Figure 2 Typical response of the Spring-Mass-Damper system.

Π On a piece of paper show how the complex roots can cancel, yielding a real the time domain solution.

Our solution to this is as follows:

Suppose, after a partial fraction expansion we obtain the following solution with conjugate complex roots:

$$\frac{1}{(s + a + bj)} + \frac{1}{(s + a - bj)}$$

Inverse Laplace transformation will yield:

$\exp(-(a + bj) . t) + \exp(-(a - bj) . t)$

This can be rewritten to:

$\exp(-a.t) . \{\exp(bj.t) + \exp(-bj.t)\}$

By means of Eulers rule ($\exp(\alpha j) = \cos(\alpha) + j.\sin(\alpha)$) we can rewrite the last result to:

$\exp(-a.t) . \{\cos(b.t) + j.\sin(b.t) + \cos(-b.t) + j.\sin(-b.t)\}$

However, as $\sin(\alpha) = -\sin(-\alpha)$ and $\cos(\alpha) = \cos(-\alpha)$, this reduces to:

$\exp(-a . t) . \{2.\cos(b.t)\} = 2. \exp(-a . t) . \cos(b . t)$, which demonstrates the cancellation of the conjugate complex poles.

Although rare in bioprocesses, this kind of behaviour can easily be introduced by closing a control loop.

Higher-order systems can be taken as the convolution of first and second-order dynamics, in which the output of one block is the input of a new block.

3 Stable and unstable systems

stable steady state

The time-domain solution of the above example contains two terms with negative exponents. When time increases, the magnitude of both exponential terms will decrease, until they can be neglected. The system reaches a steady state. The system is stable.

unstable state

If, on the other hand, one of the exponents is positive, the magnitude of the exponent would increase exponentially with time. Therefore, no steady state will be reached in such a case: the variable of interest will run completely out of hand. These situations are called unstable. A very common unstable system is the exponential growth of biomass in a batch culture. Two remarks must be made here however. First, the fact is stressed that, although uncommon in process control, for biotechnological processes, an unstable mode is sometimes required. Second, the exponential growth of biomass cannot be permanent. The exponential solution is only valid for a limited period of time.

The characteristic time(s) of a system can be found in the exponents of the time-domain solution:

$x(t) = C_3 + C_2 . \exp(-C_1.t)$

then: $T = 1/C_1$

with C_1: a positive number

Clearly, the characteristic times of a stable linear system are directly related to the poles of the system: a characteristic time is the inverse of the real part of a pole (note the possibility of complex roots of the denominator polynomial).

Whether or not a system is stable is also immediately evident from the poles of the system:

A system is stable if the real part of all poles is negative.

Again, the transfer function contains all necessary information.

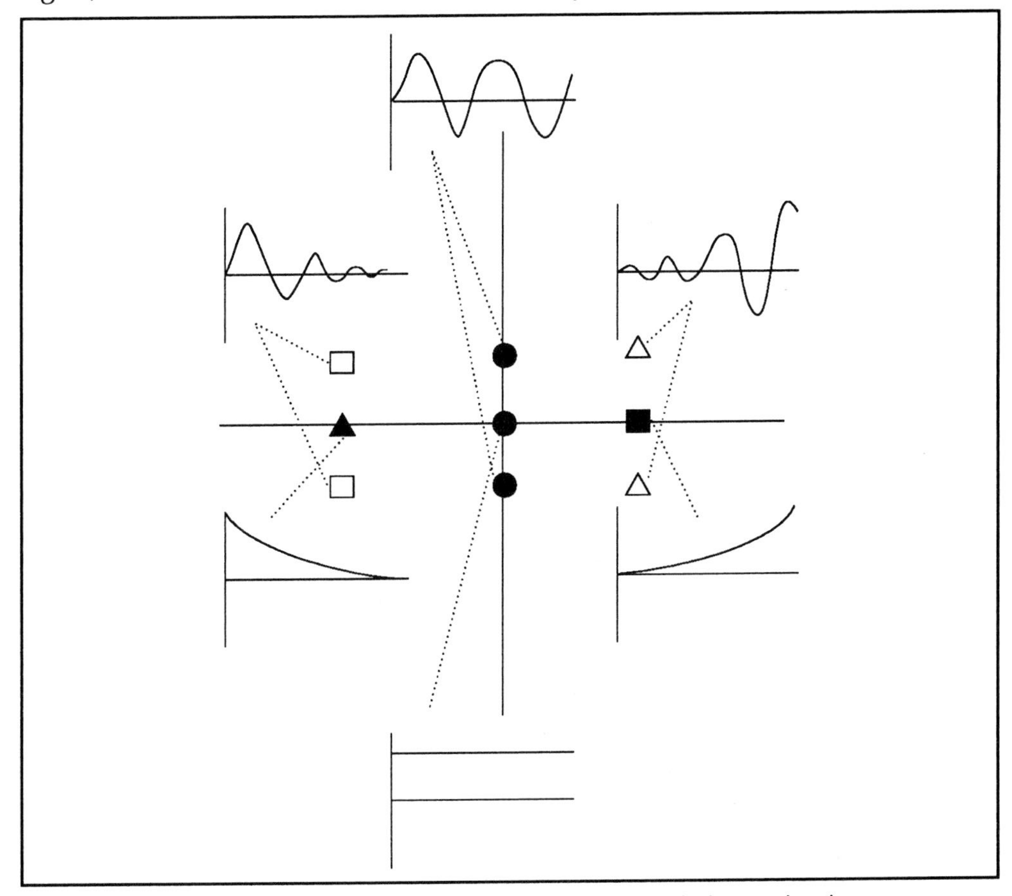

Figure 3 Relation between time-domain behaviour and location of poles in the complex plane.

As the poles of the system can be complex, their values can be made visible in a complex plane, with horizontally, the real value and vertically the imaginary value of the roots. Due to the fact that complex roots always come in conjugate pairs, the complex plane is symmetric in the horizontal axis. If all poles are in the left half plane (all real parts are negative), the system is stable (remember that even one pole in the right half plane will make the system unstable). If the poles are far in the left half plane, the system will be fast (its characteristic times will be small). The order of the system equals the number of given poles. Figure 3 summarises the relation between time domain behaviour and the location of the poles in the complex plane.

The importance of the roots of the denominator polynomial will be clear. These roots can be found by equating the denominator polynomial to zero. This important equation is denoted the characteristic equation of the system.

4 The Bode and Nyquist diagrams

In this section, we will discuss some widely used diagrams which conveniently summarise the system behaviour. The Bode diagram consists of a pair of plots, while the single Nyquist plot contains the same information as the pair of Bode plots. Both diagrams are presented in the frequency domain and can easily be obtained from the transfer function, by substitution of $j.\Omega$ for s. However, these diagrams are often based on experimental data in which case they are used for identification purposes. Both plots concern the open loop, non-controlled process.

There are two important characteristics: the amplitude ratio and the phase shift. Both are functions of the frequency. The amplitude ratio is defined as the ratio of the amplitude of a sinusoidal input signal and its corresponding output signal. Given the transfer function G(s), it is calculated as the modulus of the complex number $G(j.\Omega)$. The phase shift is defined as the apparent shift in phase of the sinusoidal input signal and the output signal. It is calculated as the argument of $G(j.\Omega)$. Suppose:

$$G(j.\Omega) = W = a + b.j$$

then:

$$\mathrm{mod}(W) = \sqrt{(a^2 + b^2)}$$

$$\arg(W) = \tan^{-1}(b/a)$$

For linear systems, the modulus and argument are no function of the magnitude of the input signal.

The Bode diagrams show:

- the behaviour of the logarithm of the amplitude ratio as a function of the frequency;
- the phase shift as a function of the frequency.

For both diagrams, the frequency is plotted on a logarithmic scale. First and second order systems possess very characteristic bode diagrams. These are given in Figure 4 and 5. Note that the phase shift for high frequencies is 90.n, where n is the order of the transfer function.

In the Nyquist diagram, the real and imaginary parts of $G(j.\Omega)$ are plotted directly, starting with Omega = 0, and continuing to a high frequency. The frequency now rises along the curve. Usually, it is not given explicitly. The modulus and argument are given implicitly by this figure. The Nyquist diagrams of the first and second order systems are given in Figures 6 and 7. For clarity, the frequency is plotted along the curves. Again, the ultimate phase shift tells the order of the system. Here, however, the ultimate phase shift is seen from the angle at which the Nyquist curve approaches the origin.

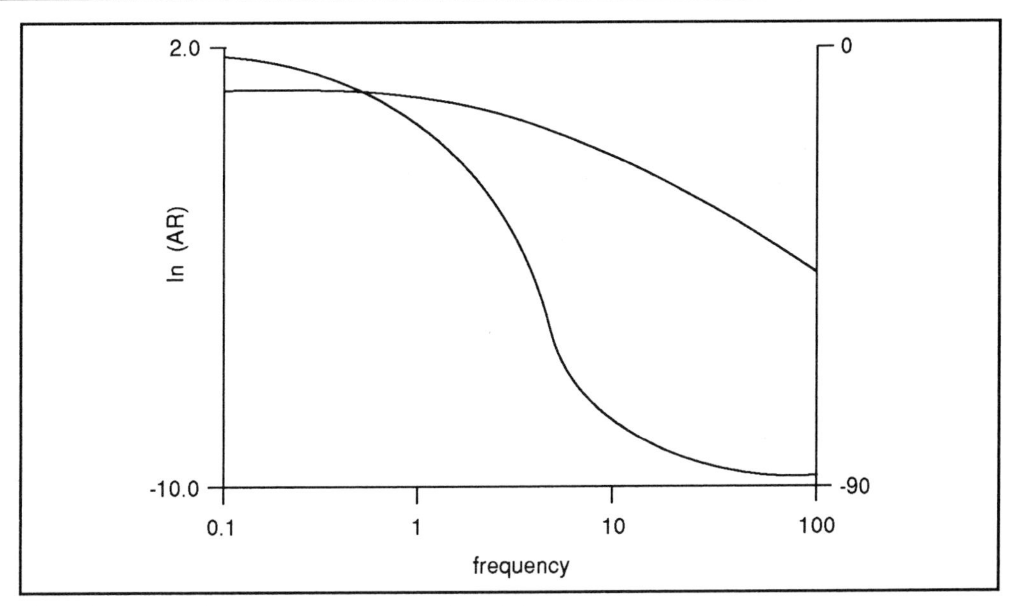

Figure 4 Bode amplitude and phase diagrams for a first order system.

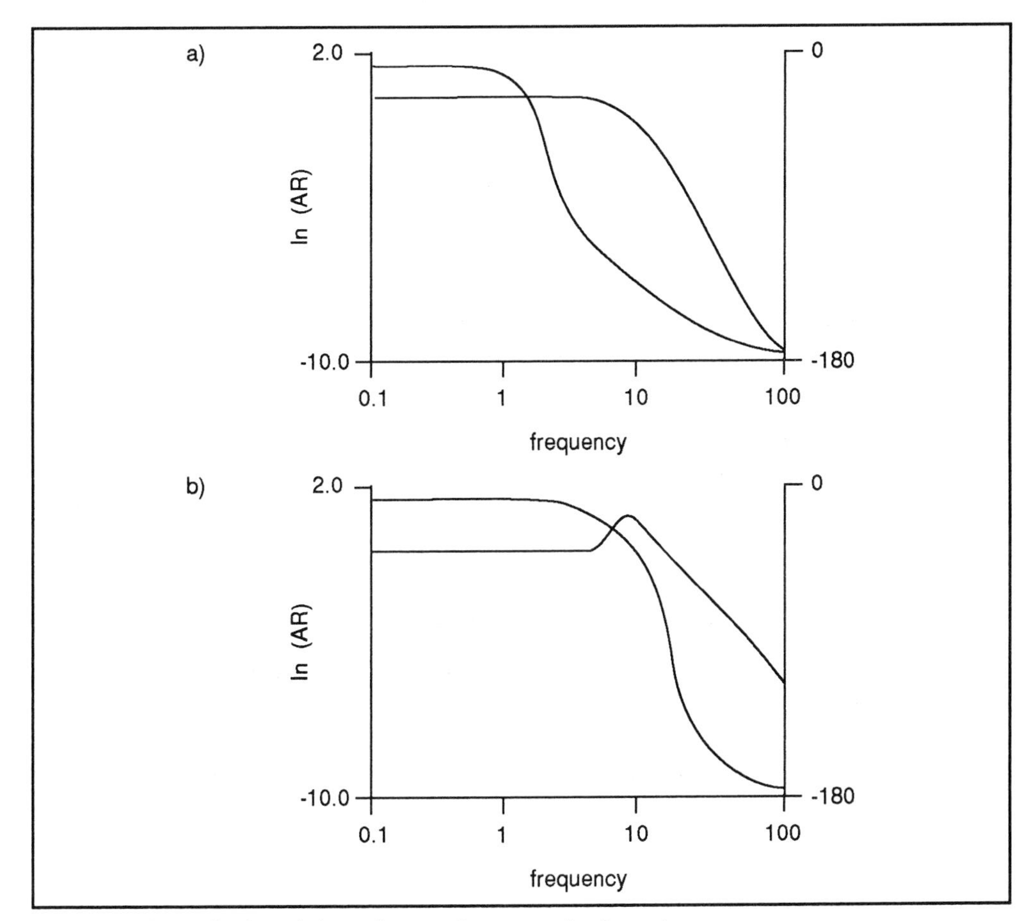

Figure 5 Bode amplitude and phase diagrams for a second order system.

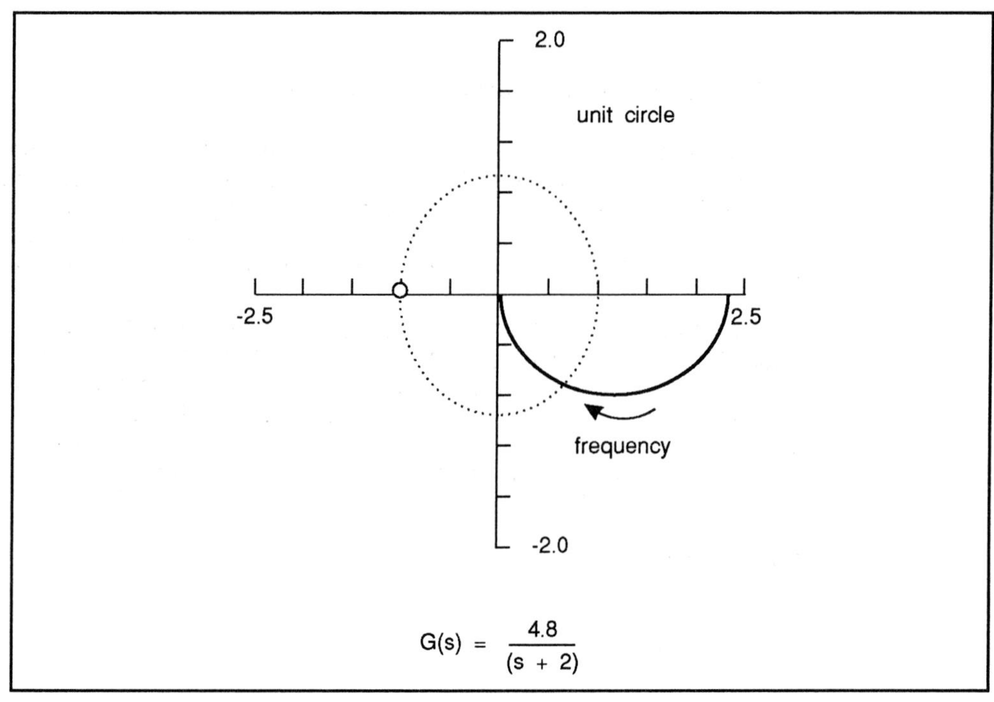

Figure 6 Nyguist diagram for a first order system.

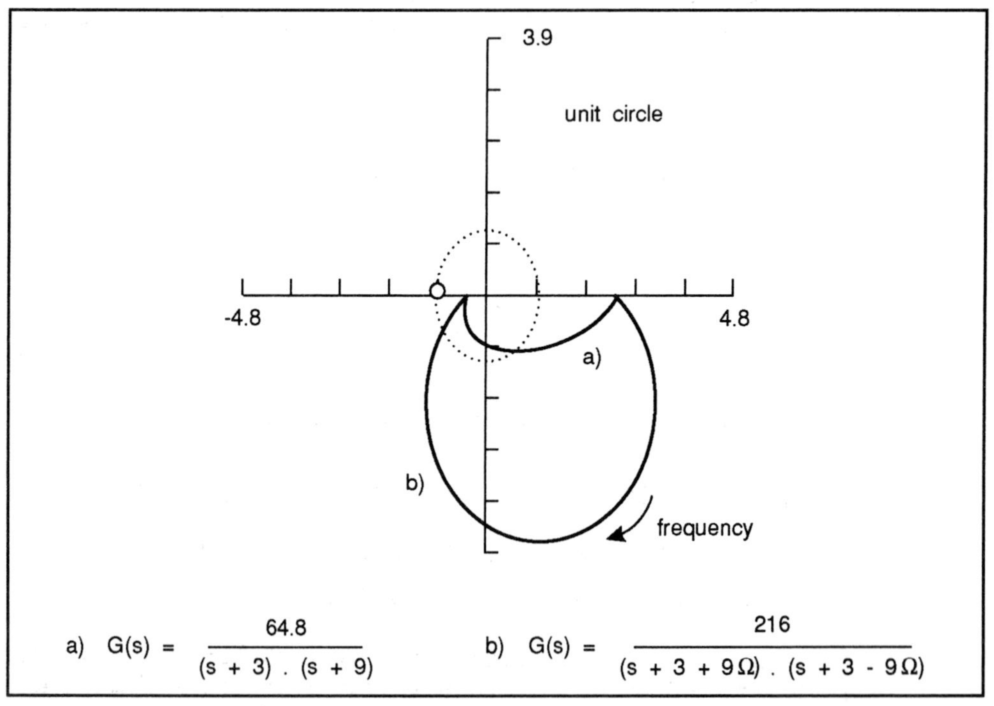

Figure 7 Nyguist diagram for a second order system.

5 P, PI and PID controller actions

In the core text, we described the principles behind P, PI and PID controllers. We learnt that 'nervous' controllers had some undesirable features. We did so in a descriptive manner. In this section we continue the type of analysis we have been following in this appendix by applying it to a PID controller, ie a controller with a differentiating element. We will then go on to discuss transfer functions in closed loop systems.

We remind you that a differentiating element can be added to a controller. As the name suggest, the differentiating element multiplies the derivative of its input signal by a constant K_d. A small but steep change of the output of a process will then cause a large controller action. This way, the controller acts even before the output has changed dramatically. Clearly, a differentiating action makes the controller more nervous. Measurement noise also has a dramatic effect on the differentiating action. Again, the D action can be combined with a P and/or I action (yielding a PD or a PID controller).

The complete formula for a PID controller is then:

$$u = C + K_p \cdot (\varepsilon + K_i \cdot \int \varepsilon\, dt + K_d \cdot d\varepsilon/dt)$$

with: ε = Setpoint - Output

The generated process input u is the output of the controller block, while the error signal ε is its input. We can therefore determine the transfer function of the controller:

$$G(s) = K_p \cdot (1 + K_i/s + K_d \cdot s)$$

Π Derive the latter transfer function yourself.

Here is our solution:

$$u = C + K_p \cdot (\varepsilon + K_i \cdot \int \varepsilon dt + K_d \cdot d\varepsilon / dt)$$

$$L[u] = L[C + K_p \cdot \varepsilon + K_p \cdot K_i \cdot \int \varepsilon dt + K_p \cdot K_d \cdot d\varepsilon / dt]$$

$$L[u] = K_p \cdot L[\varepsilon] + K_p \cdot K_i \cdot L[\int \varepsilon dt] + K_p \cdot Kd \; L[d\varepsilon/dt]$$

$$L[u] = K_p \cdot L[\varepsilon] + K_p \cdot K_i \cdot L[\varepsilon]/s + K_p \cdot K_d \cdot s \cdot L[\varepsilon]$$

$$G(s) \equiv L[u]/L[\varepsilon] = K_p \cdot (1 + K_i/s + K_d \cdot s)$$

Note that the transfer functions of simpler controllers can be obtained by equating the respective K's to zero. This transfer function is useful in determining the effect of the controller on the system behaviour, and will be used in the following sections. Preferably, the most elementary controller is used.

6 Closing the loop

So far, we have only spoken about transfer functions of open loop systems. Closing the feedback loop however has a dramatic influence on the transfer function. From the discussion in Section 5, we know the general transfer function of PID controllers, denoted as G(s). Let us assume that P(s) is the known transfer function of a process we want to control. The feedback loop is given in Figure 6. Due to the addition of the controller, the setpoint is now regarded as the input signal of the combined process, while the original measured output remains the output of the total process. As P(s) is the transfer function of the process, we already know:

$$\mathcal{L}[M(t)](s) = P(s) \,.\, \mathcal{L}[u(t)](s)$$

denoted as:

$$M(s) = P(s) \,.\, u(s)$$

We also know that:

$$u(s) = G(s) \,.\, \varepsilon(s)$$

Without closing the feedback loop, the transfer function of the open loop system is easily obtained:

$$M(s) = P(s) \,.\, G(s) \,.\, \varepsilon(s)$$

$$M(s)/\varepsilon(s) = P(s) \,.\, G(s)$$

When the loop is closed, of course, the following will hold:

$$\varepsilon(s) = S(s) - M(s)$$

Applying this result to the transfer function of the open loop system, will bring us to the transfer function of the closed loop system:

$$M(s)/S(s) = P(s) \,.\, G(s) \,/\, [1 + P(s) \,.\, G(s)]$$

(Note how easy one can work with these transfer functions.)

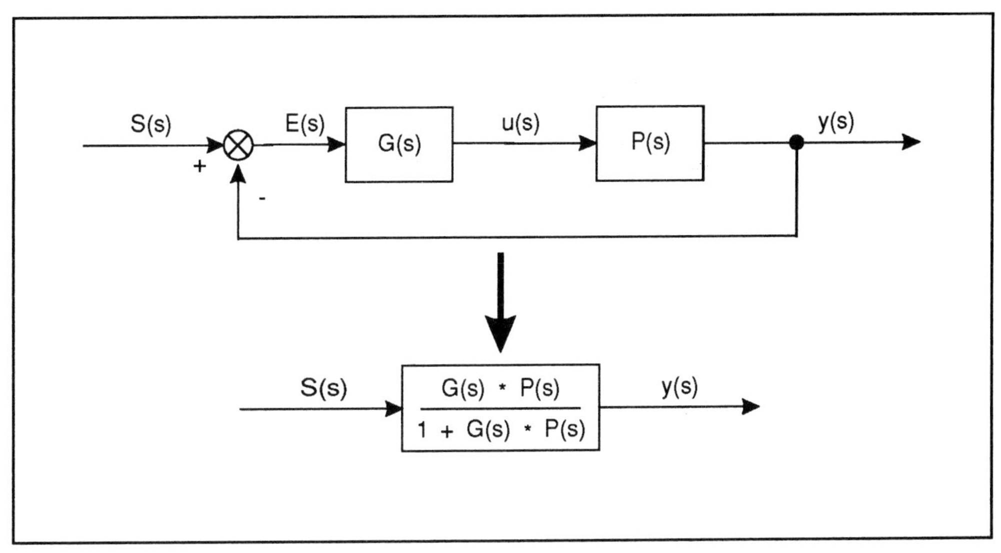

Figure 8 The open loop versus the closed loop system.

We will now take a closer look at this closed loop transfer function. First of all, we note that the denominator polynomial has changed. As a result of this, the characteristic equation has become:

$1 + P(s) . G(s) = 0$

A new set of poles will be the result. The characteristics of the feedback system may differ completely from the original system. More important, these characteristics can be influenced by the control engineer by choosing a suitable controller G(s).

Π Is the number of poles and/or zeros influenced by closing the control loop?

Let us assume that the numerator and denominator polynomials of an open-loop system are: N(s) and D(s). Then the transfer function of the closed loop system will become:

$\{N(s)/D(s) / \{1 + N(s)/D(s)\}$

This can easily be rewritten to:

$\{N(s)\} / \{N(s) + D(s)\}$

As, for real processes, the order of the numerator polynomial is never higher than the order of the denominator polynomial, it will be clear that the number of poles and zeros do not change by closing the loop.

Note that, as a controller has it own transfer function, this certainly will effect the order of the closed loop system: A P-action does not alter the number of poles or zeros, a D-action adds one zero and a I-action adds one zero and one pole (in the origin).

6.1 The Nyquist and Bode criteria

Using the Bode or Nyquist diagrams of the open loop system, we can study the behaviour of a closed loop system. Here, we will study the stability of the closed loop system. For each of the diagrams, simple rules of thumb exists, that immediately tell whether or not a closed loop system will be stable.

The following simple reasoning is necessary to understand these rules. Suppose we apply an input signal of amplitude 1 with a certain frequency. Let us furthermore assume that, for the given frequency, the resulting output signal has an amplitude which is larger than 1. If the output signal is negative, while the input has a positive value, the input signal will be increased due to a negative feedback loop. Clearly, the system will then be unstable. If the output has a negative extreme, while the input has a positive extreme, there is a phase shift of 180 degrees. Therefore, the rule holds that if a system generates, for any frequency, an output signal whose magnitude is larger than its input signal at a phase shift of 180 degrees, the system will become unstable if the loop is closed.

Another way of obtaining the same insight is to look at the open and closed loop transfer functions:

open loop: $G(s)$

closed loop: $G(s)/\{1 + G(s)\}$

In order to be stable, the denominator may not become zero for any frequency. Therefore, $G(s)$, the open loop transfer function may not become -1. This means that at a phase shift of 180 degrees, the amplification must remain below 1.

For both plots, the criterion is easily verified: for the Bode diagram, we look at the location where the phase shift is 180 degrees, then, the amplitude ratio must be smaller than one. For the Nyquist diagram, the rule holds that the curve of the open loop system must pass the point (-1,0) at the right side.

Two margins are defined: the gain margin and the phase margin. The gain margin is defined as $1/M$, where M is the amplitude ratio of the frequency with a phase shift of 180 degrees. The phase margin is defined as 180-P, where P is the phase shift at a frequency where the amplitude ratio is 1. Clearly, a gain margin equal to or below 1 yields an unstable control system, while a phase margin of zero will have the same effect. For most practical situations, the control engineer applies a minimum gain margin of 1.7 and a minimum phase margin of 30 degrees. Then, only severe model variations, uncertainties, disturbances etc can make the system unstable.

Figure 9 shows the minimal design margins for the Bode diagrams, while Figure 10 shows the margins for the Nyquist diagrams.

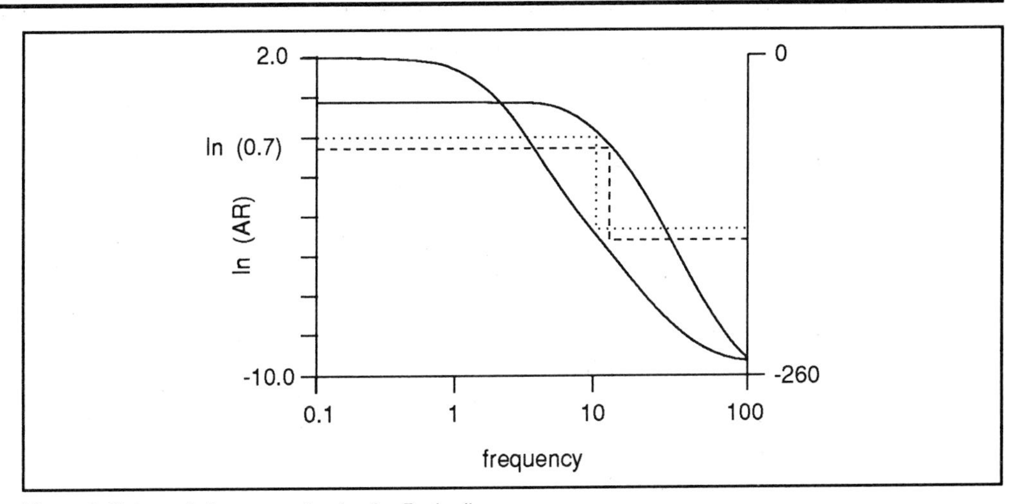

Figure 9 Gain and phase margins for the Bode diagrams.

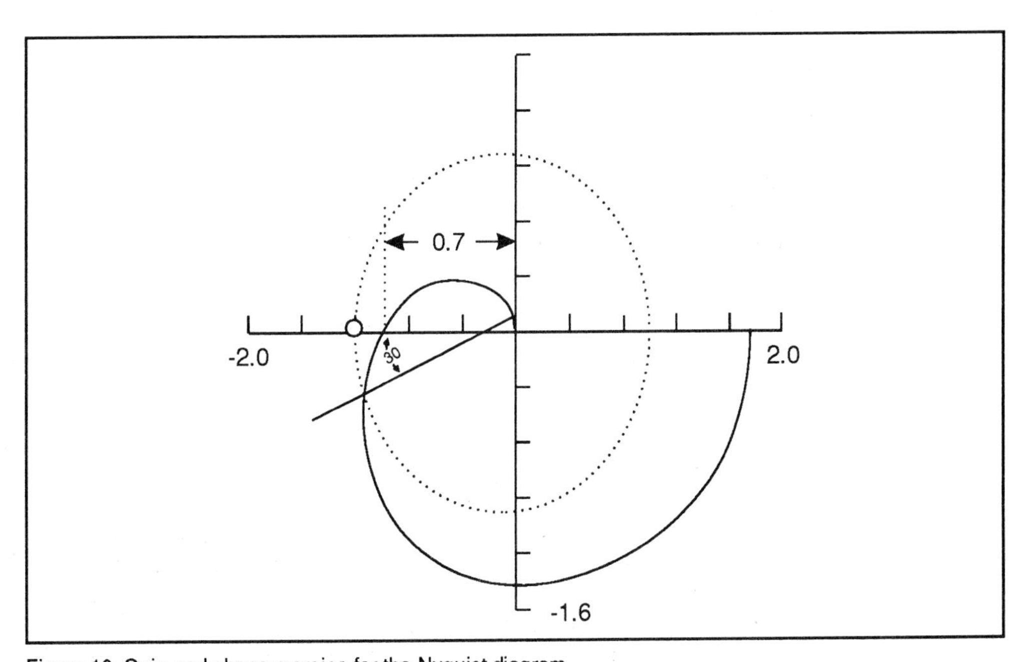

Figure 10 Gain and phase margins for the Nyguist diagram.

The control engineer not only uses these plots to investigate stability. Because the influence of each of the control actions can easily be determined, these plots are often used to decide which kind of controller is necessary for the given process.

6.2 Tuning PID-type controllers

The margins defined above can effectively be used to tune a controller. However, in this section, we will discuss two other very common rules of thumb for tuning PID-type controllers. Besides the demand of the controlled (sub)system to be stable, we may have some other demands: settling time, maximal steady state error, sensitivity to measurement noise, etc.

6.2.1 Cohen-Coon rules

Cohen-Coon rules

The first method discussed here will be the Cohen-Coon rules. Cohen-Coon based their rules on an approximate model of the process under consideration. This approximate model has a the following transfer function:

$$G(s) = K \,.\, \exp(-t_d.s) \,/\, (Ts + 1)$$

This transfer function describes a first order process with a dead-time (which is described by the exponential term in the transfer function). The dead-time is equal to t_d. An arbitrarily process is now approximated with the above transfer function. For many processes, this is possible. K, t_d and T are now found according to Figure 11. Once K, t_d and T are found, the controller can be tuned according to Table 2.

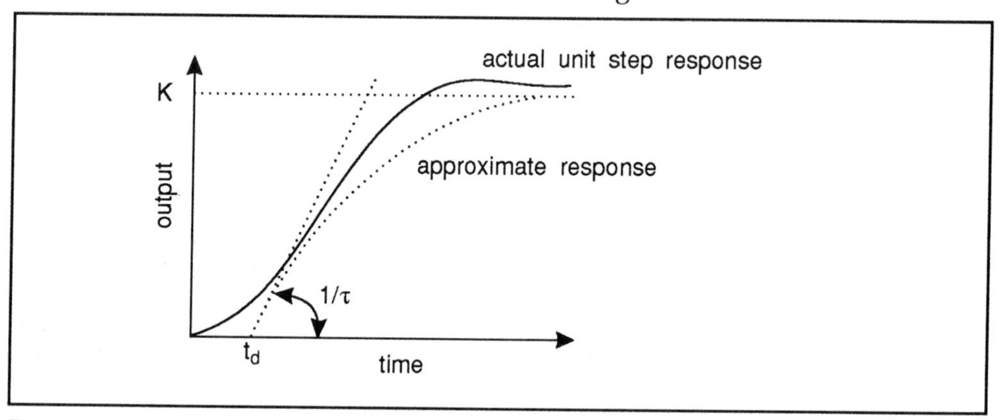

Figure 11 Relation between model parameters and the time response of a step change on the process input.

P:	$K_p = (1/K) \,.\, (T/t_d) \,.\, (1.0 + t_d/3T)$
PI:	$K_p = (1/K) \,.\, (T/t_d) \,.\, (0.9 + t_d/12T)$ $T_i = t_d \,.\, (30 + 3t_d/T)/(9 + 20\, t_d/T)$
PID:	$K_p = (1/K) \,.\, (T/t_d) \,.\, (3/4 + t_d/4T)$ $T_i = t_d \,.\, (32 + 6t_d/T)/(13 + 8t_d/T)$ $T_d = t_d \,.\, 4/(11 + 2t_d/T)$

Table 2 Cohen-Coon controller settings.

6.2.2 Ziegler-Nichols rules

Ziegler-Nichols rules

The Ziegler-Nichols rules form a different approach. In an experiment, the maximal proportional feedback gain is found. The idea is that this gain is related to the optimal gain, as are the integrating and differentiating actions to the oscillating period. The philosophy compares well with that of the gain and phase margins from the previous section. Some simple experiments have to be done first. In short, the method works as follows:

- Bring the process to the desired working point.
- Close the feedback loop. Use a proportional controller with a variable, but initially low, K_p. Increase K_p until the system starts to oscillate. Note that the controlled system has now reached its stability limit.
- The K_p for which this occurs is called the ultimate gain K_u, while the ultimate period T_u is the period of the oscillations. The Ziegler-Nichols settings corresponding to K_u and T_u are given in Table 3.

Controller type	K_p	T_i	t_d
P	0.50 K_u	-	-
PI	0.45 K_u	T_u/1.2	-
PID	0.60 K_u	T_u/2.0	T_u/8.0

Table 3 Ziegler-Nichols controller settings.

Note that the optimal K_p for the P controller is only 0.5 times the maximal gain K_u (a gain margin of 2). Note also that, in Table 2, the proportional factor for a PI controller is somewhat lower, and for a PID controller somewhat higher than for a simple P controller. This indicates that an integrating action deteriorates stability (the proportional factor is more moderate), while a D action generally improves stability (due to its predictive property).

It will be clear that both methods are more or less empirical rules. For individual situations, one may still have to adjust the controller settings in order to obtain an optimal result.

6.2.3 The root-locus method

The root-locus method for tuning a controller is completely based on the transfer function. Again, we have to decide a priori which type of controller is to be used. Then, the transfer function of the controlled system is determined. Here however, it is essential that the closed loop transfer function is taken as a function of the proportional gain factor K_p. The poles and zeros of the closed loop system are then also functions of the K_p. In principle, the root locus method uses a plot of the location (values) of the poles and zeros (in the complex plane) as a function of the proportional gain factor K_p.

In order to obtain a satisfying process behaviour, the control engineer places the most dominant poles of the controlled system at an a priori chosen location. This is called pole placement.

Π Which are the most dominant poles in general?

The right-most poles. If any pole is in the right half plane, this single pole will cause the signal to exponentially grow with time. If all poles are in the left half plane (the system is stable), the right-most poles represent the slowest time characteristics. Therefore, while the effect of this pole is evident for the longest time.

Besides that it facilitates pole placement, the root locus method offers a valuable insight in the stability of the system, by determining in what conditions the root loci enter the right half plane. By applying certain rules to predict the curves of the root locus from the root loci at $K_p = 0$, the control engineer can determine what kind of controller is needed in order to get the poles at the desired location and to obtain a robust controller.

7 Non-linear control

The Laplace transform and other above mentioned techniques for system analysis are in principal only useful for linear systems. For non-linear systems, the principle of superposition doesn't hold. This means that, for instance, if one input signal results in one type of behaviour, twice that input signal may yield quite a different behaviour. Linear systems can only possess one steady state point, while non-linear systems sometimes have more than one. Which of those will eventually be reached depends on the starting conditions. In general, non-linear systems can be much more complicated than linear systems.

A well known biotechnological example is the continuous culture of two microorganisms, say A and B. Organism A is capable of growing fastest at high substrate concentrations, while organism B has a higher growth rate at low substrate concentrations due to a higher affinity to the substrate. Which of the two will eventually survive depends on the dilution rate that is applied.

In practice, most (biotechnological) processes are non-linear. In this chapter, we will therefore discuss some methods to address these non-linear systems. It must be noted however, that no general approach exists for non-linear systems. The approaches discussed in this chapter will not always be successful.

7.1 Linearisation

One efficient approach for many non-linear systems, and no doubt the most important, is linearisation. In this way, a linear approximation of the original non-linear system can be obtained, so the tools for linear systems become applicable.

As linear systems can only have one steady state point in the origin, linearisation of the non-linear differential equations normally requires that we must first define the steady state point as the origin. This can be done conveniently by using perturbation variables. A perturbation variable contains the difference between the real value and its value in steady state. Let us consider the following non-linear example:

$$dX/dt = - D \,.\, X + q_{max} \,.\, S/(K_S + S) \,.\, Y$$

$$dS/dt = D\,(S_{in} - S) - q_{max} \,.\, S/(K_S + S)$$

Here, X and S are assumed to be state variables, while D and S_{in} are considered process inputs (our handles to control the process). The perturbation variable become:

$x_1 = X - X^*$

$x_2 = S - S^*$

where (S^*, X^*) is the steady state point, which belong to a certain steady state input (D^*, S_{in}^*). The input variables are also transformed to the origin:

$U_1 = D - D^*$

$U_2 = S_{in} - S_{in}^*$

Next step in the linearisation procedure is to calculate the partial derivatives of dX/dt and dS/dt to each of the state and input variables at the steady state point (in our example, this will therefore yield 8 partial derivatives: each equation to the four variables X, S, D and S_{in}):

the first differential equation:

$$d(dX/dt)/dX \Big|_{(S^*, X^*, D^*, S_{in}^*)} = -D^*$$

$$d(dX/dt)/dS \Big|_{(S^*, X^*, D^*, S_{in}^*)} = Y \, . \, q_{max} \, . \, Ks/(Ks + S^*)^2$$

$$d(dX/dt)/dD \Big|_{(S^*, X^*, D^*, S_{in}^*)} = -X^*$$

$$d(dX/dt)/dS_{in} \Big|_{(S^*, X^*, D^*, S_{in}^*)} = 0$$

Π Try to obtain the solution for the second differential equation yourself. The solution can be found in the feedback section.

Your answer should be:

$$d(dS/dt)/dX \Big|_{(S^*, X^*, D^*, S_{in}^*)} = 0$$

$$d(dS/dt)/dS \Big|_{(S^*, X^*, D^*, S_{in}^*)} = -D^* - q_{max} \, . \, Ks/(Ks + S^*)^2$$

$$d(dS/dt)/dD \Big|_{(S^*, X^*, D^*, S_{in}^*)} = S_{in}^*$$

$$d(dS/dt)/dS_{in} \Big|_{(S^*, X^*, D^*, S_{in}^*)} = D^*$$

The linear system then look as follows:

$$\underline{\dot{x}} + A \, \underline{x} + B \, \underline{u}$$

$\underline{x} = (x_1, x_2)$

$\underline{u} = (u_1, u_2)$

$$A = \begin{vmatrix} d(dX/dt)/dX & d(dX/dt)/dS \\ d(dS/dt)/dX & d(dS/dt)/dS \end{vmatrix}$$

$$B = \begin{vmatrix} d(dX/dt)/dD & d(dX/dt)/dS_{in} \\ d(dS/dt)/dD & d(dS/dt)/dS_{in} \end{vmatrix}$$

The system from above is a linear (MIMO) system. One can now study the transfer function of, for instance, input D to output X by means of linear system analysis.

7.2 Simulation as a tool

Whenever a process model is available, simulation can be used for controller development and testing. As there are less general methods available for non-linear systems, simulation is usually applied for these processes. There are two major advantages: simulating is, of course, much cheaper (and faster) then performing tests with the actual process, almost no limitations exist with respect to simulation. All kinds of non-linear behaviour can therefore be investigated. However, there are also some disadvantages. The most obvious disadvantages are:

- a mathematical process model must be available;
- the model may not be accurate enough or some model parameters may be unknown (resulting in unreliable results);
- a lot of simulations usually have to be carried out, in order to find out how the system responds in different situations;
- some simulations take a long time (the model may be too complex).

Π Why is it not necessary to carry out much simulations for linear systems?

For linear systems, the very strong principle of superposition holds. Therefore, if one step-response is known, the response to any input signal can be calculated by proper multiplication and summation of the step response. In contrast to non-linear systems, linear systems can posses just one type of behaviour.

A practical approach is to use a simplified process model for the development of the controller. Simulations will then take less time, and less simulations have to be carried out (again due to the simplicity of the model). Finally, the controller is tested by applying it to the process complex model. If necessary, some parameters and starting conditions of the process may then be altered in order to investigate the robustness of the controller. If, for some reason, the controller is not satisfying, one should try to use a less simplified model and restart the procedure.

As linearisation is some kind of model simplification, controller design based on a linearised process model should also be tested by means of simulating the original non-linear model, before applying it to the real process.

Π What is the advantage of linearisation if simulation is still necessary?

After linearisation, more general methods become available for the controller development. Therefore, the controller development will usually be faster, while at least for the linear system, the controller will be reliable.

7.3 State space method

As mentioned in the previous section, a lot of simulations are necessary in order to investigate the system behaviour for different situations. The state space method is used to determine which situations are to be investigated. However, the visual method is, in its most simple form, limited to processes with two state variables. The method is based on the fact that trajectories cannot cross in the state space.

Π Why is this so?

For a trajectory to meet another trajectory, the system state must become equal to that of the other trajectory. As the system state determines completely the future system behaviour, the trajectories must follow the same path from the meeting point on, hence, a crossing point cannot exist.

Therefore, one well chosen, single simulation can effectively determine a major part of the system behaviour. First, one determines all steady state working points. This can be done by equating the differentials to zero and solving the non-linear equations. These points are plotted in a two dimensional figure with one state variable on one axis and the other state variable on the other axis. Some processes may have several of these points. Secondly, one simulates the process behaviour, starting from a point on some distance of the steady state point that is investigated. The time domain simulation results are then plotted in the state space (note that the time aspect will get lost).

There are two possibilities:

- the state space curve reaches (eventually) the steady state point;
- it deviates from the steady state point and goes to infinity (or to one of the other working points).

In the first situation, the starting point of the simulation is said to belong to the domain of attraction of the steady state point. Note that all points on the state space curve belong to the domain of attraction. A curve of the second situation gives points outside the domain of attraction. Note that this curve may still be part of the domain of attraction of another steady state point.

Π How does the state space look for linear systems?

Linear systems can only have one steady state point (in the origin). If the system is stable, all curves will go towards the origin. The region of attraction is then the whole state space. The curves may all spirulate towards the origin in the case of complex poles (again, without crossing one another), or they may all go more or less straight (real poles). If the system is unstable, all curves will move from the origin. Then, no the region of attraction exists.

Simulations (including the controller) are applied to find the domain of attraction for each of the steady state working points. One therefore concentrates on the borders of the domain of attraction, yielding suitable starting conditions for the simulations. A robust controller will generate a large domain of attraction for the preferred working point. A small domain of attraction demonstrates the close possibility of instable or undesirable system behaviour.

To illustrate the concept, let us consider the following example. Bacterium strain X is grown in continuous culture. It is assumed that the process can be described by the following equations:

$$dX/dt = -D.X + \mu_{max} . S/(KS + S) . K_I/(K_I + S) . X$$

$$dS/dt = D (S_{in} - S) - (\mu_{max} / Y_{sx}) . S/(K_S + S) . K_I/(K_I + S) . X - M_S . X$$

The process is characterised by substrate inhibition at high substrate concentrations (described by the term with K_I), and substrate limitation (the Monod term) at low substrate concentrations. Of course: $K_I >> K_S$. The Pirt concept is used to describe maintenance requirements.

First, we find the steady state working points, by equating the differentials to zero. After some manipulation, this will yield the following three solutions:

$$S_0^* = S_{in}^*$$

$$X_0^* = 0$$

$$S_1^* = K_I . \mu_{max} - D^* . K_S - D^* . K_I + \text{root} [(D^* . K_S + D^* . K_I - \mu_{max} . K_r)2 -4 . D^{*2} . K_S . K_r)]\} / \{2 . D^*\}$$

$$X_1^* = D^* . (S_{in}^* - S_1^*) /(D^* / Y_{sx} + M_S)$$

$$S_2^* = \{K_I . \mu_{max} -D^* . KS - D^* . Kr + = - \text{root} [(D^* . K_S + D^* . K_I - \mu_{max} . K_r) 2 -4 . D^{*2} . K_S . K_r)]\} / \{2 . D^*\}$$

These are the three steady state working points. Besides the trivial solution (S_0^*, X_0^*), the growth rate may equal the dilution rate at two distinct points. One at a low substrate concentration (substrate limitation), and one at a high substrate concentration (substrate inhibition). Suppose we have the steady state at high substrate concentration (substrate inhibition). Due to a minor disturbance, S decreases temporarily. This will cause the growth rate to increase (less inhibition) and therefore, S will decrease further. Clearly, the solution at high substrate concentration is unstable.

Π What happens if S increases due to a disturbance?

Then, the substrate inhibition will deteriorate. Therefore, the growth rate will decrease, resulting in a reduced substrate uptake rate, which, in turn, results in a further increase of the substrate concentration: again, the trajectory moves away from the steady state point (however, in the opposite direction).

Figure 12 gives the system behaviour in the state space. As may be clear from this figure, points (S_0^*, X_0^*) and (S_2^*, X_2^*) are stable. Each possesses its own region of attraction. Point (S_1^*, X_1^*) is unstable. It does not have any region of attraction (it rejects trajectories beginning in its neighbourhood). Note that the time aspect of the simulations has disappeared. In spite of what some short trajectories may suggest, simulations, staring in the neighbourhood of (S_1^*, X_1^*) will take a long time.

Note that the given example does not include a controller. A suitable controller may relocate the steady state working point, while at the same time, the region of attraction is enlarged.

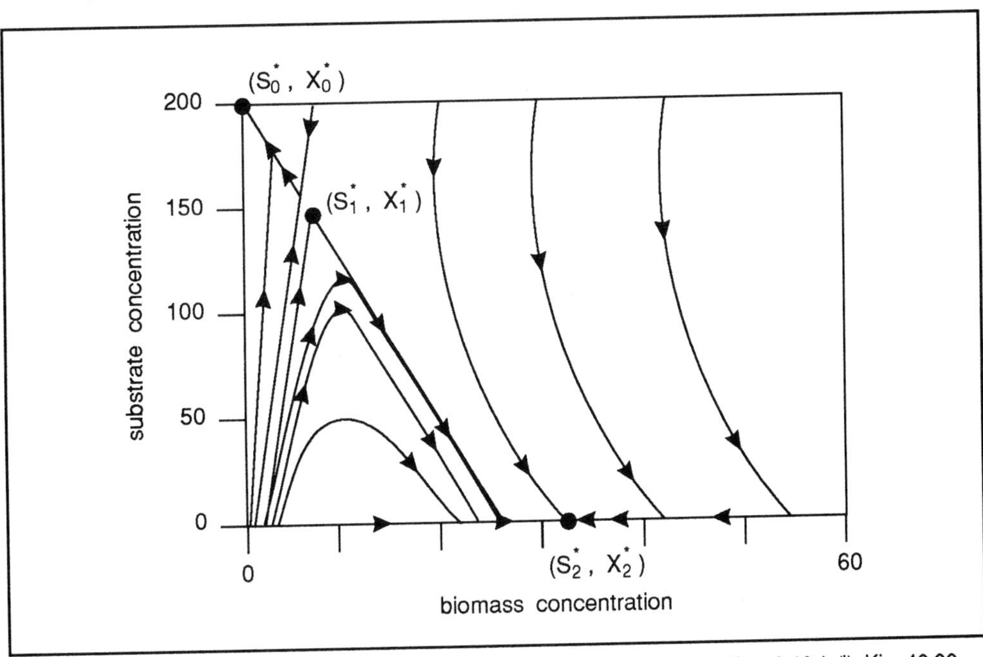

Figure 12 State space behaviour of the non-linear process. Parameter values: Ks = 0.10 (g/l), Ki = 40.00 (g/l), μ_{max} = 0.20 (1/h), Y_{sx} = 0.20 (-), Ms = 0.05 (1/h), S_{in} = 200.00 (g/l), D = 0.04 (1/h). See text.

8 State estimation

In this section we extend the discussion of state observers and estimators described in the main text. You will recall that state observers are used to reconstruct the system state including estimation of non-measurable variables. The example we gave in the chapter text was estimating substrate concentration from product (ethanol) concentration (see Section 9.5).

8.1 Structure of observers

Observers are state estimators that are based on a state space model of the process (model based state estimator). They consist of two parts. Firstly, a model is used to produce an estimation of the system state on the basis of a real-time simulation. Secondly, the difference between the real measured values (v) and the expected values (estimated v) is used to correct the state estimates. As in our previous example, the model produces an estimate for the substrate concentration, as well as an estimate for the actual measurement: the ethanol concentration. The difference between the expected ethanol concentration and the measured value is used to correct both estimates (for substrate and for ethanol). In this way, the observer uses both calculated data from the model and actual measurement information of the process under consideration.

If the following state space description is given (subscript p denotes that this particular model of the process is an exact description of the process):

$$\dot{\underline{x}} = \underline{f}_p(\underline{x}, \underline{u}) + \underline{w}(t)$$

$$\underline{y} = \underline{g}_p(\underline{x}) + \underline{v}(t)$$

with:

$\dot{\underline{x}}$: vector of state variables of the process (eg biomass, substrate and product concentrations);

$\underline{u}$: vector of inputs of the process (eg the dilution rate and the influent substrate concentration)

$\underline{f}_p$: vector of non-linear differential equations

$\underline{w}(t)$: vector of measurements taken from the process (eg the product concentration)

g_p: vector of relations between $\underline{y}$ and $\underline{x}$ (often $\underline{y}$ is just a subset of $\underline{x}$)

$\underline{v}(t)$: measurement noise vector

then the observer equation will become:

$$\dot{\hat{\underline{x}}} = \underline{f}_m(\hat{\underline{x}}, \underline{u}) + L(\underline{y} - g_m(\hat{\underline{x}}))$$

with:

$\underline{f}_m$: model approximation for $\underline{f}_p$ (not exact)

$\underline{x}$: vector of estimates for $\underline{x}$

$g_m(\hat{\underline{x}})$: expected value for $\underline{y}$

L: observer gain matrix

The relative weighing of the model and correction part can be chosen by the designer by selecting the gain matrix (observer matrix) L.

If, for example, the model is very reliable and the measurements are noisy, matrix L will be chosen 'small'. This will yield state estimates close to the values that would have been generated by pure simulation. A 'large' L will yield estimates for y close to the measured values, and will therefore ignore the model to some extent. A large L can give rather noisy state estimates (if the measurements are noisy).

Note that the terms 'small' and 'large' are unsuitable for a matrix. Here, a 'small' matrix L means that L is chose so as to make eigen values of the matrix (A-LC) small (but negative).

The size of L will also effect the quality of the state reconstruction (the estimation of non-measured parts of the system state x).

8.2 Stability of the estimator

A stable state estimator will yield non-diverging state estimates. This means that the estimation error does not increase continuously, but remains limited. We will therefore first derive the error system, ie the dynamical description for the estimation errors. It can be used to investigate the propagation of the estimation error with time. The estimation error is normally denoted as $\underline{\bar{x}}$ and is simply defined as:

$$\underline{\bar{x}} = \underline{x} - \underline{\hat{x}}$$

A set of differential equations for $\underline{\bar{x}}$ can then be obtained by substrating the differential equation for $\underline{x}$ and $\underline{\bar{x}}$, thus:

$$\underline{\dot{\bar{x}}} = \underline{\dot{x}} - \underline{\hat{x}}$$
$$= \underline{f}_p(\underline{x}, \underline{u}) + \underline{w}(t) - \underline{f}_m(\underline{\hat{x}}, \underline{u}) - L[\underline{g}_p(\underline{x}) + \underline{v}(t) - \underline{g}_m(\underline{\hat{x}})] \qquad \text{(E - 8.2.1)}$$

This equation describes the dynamic behaviour of the estimation error. In practice, it is common to simplify this equation by assuming that the model is correct ($\underline{f}_m = \underline{f}_p$ and $\underline{g}_m = \underline{g}_p$). If linear differential equation are being used:

$$\underline{f}(\underline{x}, \underline{u}) = A\underline{x} + B\underline{u} + \underline{w}(t) \qquad \text{(E - 8.2.2)}$$

$$\underline{g}(\underline{x}) = C\underline{x} + \underline{v}(t)$$

the error system simplifies to the following basic form which has been used widely in the literature:

$$\underline{x} = (A - LC)\,\underline{\bar{x}} + \underline{w}(t) - L\underline{v}(t) \qquad \text{(E - 8.2.3)}$$

When dealing with non-linear equations, it is common practice to linearise these equations around an operating point.

As can be seen, the dynamic description of the estimation error $\underline{\bar{x}}$ is now an explicit function of $\underline{\bar{x}}$ itself and the noise sequences $\underline{w}$ (t) and $\underline{v}$ (t). If the noise is neglected, the error dynamics are completely determined by matrix (A-LC). Whether or not the observer is stable, is therefore determined by the poles of the linear system given by equation 3, ie the eigenvalues of matrix (A-LC) must all be in the left half plane.

8.3 Error sources

The ideal state estimator produces accurate values for all measured and non-measured state variables at all times. In practice, this is not really possible because of several problems. Firstly, the models used in the observer are often not very accurate. Unknown system disturbances may have an impact on the system state (system noise), parameters may partly be unknown or may vary with time, and even the model structure may be to some extent inadequate. Secondly, errors in the initial conditions for the observer ($\underline{x}$ at $t = 0$) will propagate with time and will therefore cause estimation errors (at least temporary). Finally, the measurements usually contain noise which causes deviations in the estimates.

To overcome most of these problems, the state estimator should be tuned in such a way that:

- errors due to incorrect initial conditions for the observer are rapidly eliminated;
- the estimates are insensitive to measurement noise;
- the estimation algorithm is not sensitive to errors in the model.

These three properties are to some extent in conflict with each other, and it is impossible to fulfil all the demands at the same time.

It was shown in the previous section, gain matrix L has a profound influence on the error dynamics. Under certain conditions, the system ($\underline{x} = (A - LC)\,\underline{x}$) can be made arbitrarily fast by a proper choice of matrix L. The larger L, the faster the estimation error will approach zero (and the faster the effect of initial errors will disappear with time). However, as is demonstrated by equation 3, if L is large, the estimation error becomes very sensitive to measurement noise $\underline{v}$ (t).

Clearly, the selection of an optimal gain matrix L is important. This can be done by means of a technique developed Kalman and Bucy in the early 1960s and known as the Kalman filter theory.

An alternative approach is to select a gain matrix L on the basis of pole placement. The philosophy behind pole placement is that the error dynamics need not be faster than 10 times the speed of the original system. Let us assume that the original system (from Equation 8.2.2) has certain characteristic times α_i. (These characteristic times can be evaluated by calculating the eigenvalues of matrix A.). The error dynamics must then have characteristic times of about $\alpha_i/10$. (The eigenvalues of matrix (A-LC) must then be 10 times those of matrix A.). In most cases, however, this will yield noisy state estimates due to measurement noise $\underline{w}$ (t). A practical choice is to select eigenvalues of matrix (A-LC) which are 2 to 4 times those of matrix A.

In deriving Equation 8.2.3, we assumed that the model was accurate. If this assumption is discarded, it can easily be shown that the state estimates may become biassed (their values may constantly be too high or too low). If, as in most biotechnological cases, a crude model is used, the bias effect requires special attention. In general, the effects of model errors can be investigated by simulation, using the (non-linear) error dynamics (Equation 8.2.1).

9 Process optimisation

In the main text we described in outline the process of process optimisation either through trial experiments or by modelling. We remind you that usually, the optimal working point or controller setpoints will already be known from laboratory experiments or previous process runs. However, one sometimes may have to calculate a suitable (preferably optimal) working point from a given process model. Although the model will not be exact, the calculated point can at least serve as a starting point. One may furthermore be confronted with a dynamic optimisation problem (eg an optimal feeding strategy for a fed-batch culture, or, an optimal temperature trajectory for penicillin production). For some situations, it will be necessary to develop an on-line optimising control strategy. Control and optimisation will then go hand in hand.

9.1 Static Optimisation

Let us consider the following dynamic system:

$$dX/dt = -D.X + \mu_e . X + Y_{sx} . R_{smax} . S/(KS + S) . (1-P/P_{max}) . X$$

$$dS/dt = D . (S_{in} - S) - R_{smax} . S/(K_S + S) . (1-P/P_{max}) . X$$

$$dP/dt = -D . P + Y_{se} . R_{smax} . S/(K_S + S) . (1-P/P_{max}) . X$$

Let us furthermore assume the following definition of the performance function of the system:

$$\text{Performance} = D . (6 . P - S_{in})$$

demonstrating the fact that the price for the product is six times the price of the substrate. Because we are interested in steady state conditions only, we can use the steady state solution for this system, which can be obtained by equating the differentials to zero. This yields a quadratic equation for S. Therefore, S is given by means of the variables, a, b and c:

$$S^* = (b + \sqrt{b^2 + 4 . a . c}) / (2 . a)$$

$$P^* = Y_{se} . (S_{in}{}^* - S^*)$$

$$X^* = D^* . (S_{in}{}^* - S^*) / \{R_{smax} . S^* / (K_s + S^*) . (1 - P^* / P_{max})\}$$

with $a = R_{smax} . Y_{sx} . Y_{se}$, $b = P_{max} . (D^* + \mu_e) + R_{smax} . Y_{sx} . Y_{se} . S_{in}{}^* - R_{smax} . Y_{sx} . P_{max}$, $c = P_{max} . (D^* + \mu_e) . K_s$

Π In fact there are three solution for the steady state. Why should we take the solution from above?

The answer to this is that the second solution is trivial: $S^* = S_{in}{}^*$ and $X^* = P^* = 0$, while in the third solution, $S^* = (b - \sqrt{b^2 + 4 . a . c}) / (2 . a)$, S^* is negative and therefore physically unrealistic.

The problem is to find the optimal dilution rate and influent substrate concentration for the given process. Figure 13 shows the performance as a function of D^* and S_{in}^* by choosing suitable starting points and step-sizes.

If no analytical solution can be found for the steady state of a complex process, one may also solve the equations numerically, by means of an iterative optimisation method. One then searches for a set of state variables for which all the derivatives are zero (an optimisation within an optimisation).

Note that the economic performance function from the above example is rather simplistic. Process economics depends on much more factors, like equipment investments, market value of product and substrate, downstream processing costs, etc. This would yield rather complex performance functions, and moreover, many parameters of these functions are unknown or variable with time. Therefore, one often uses a performance function which is more directly related to the biological conversion step itself (like maximising the growth rate, production rate, enzyme stability, etc). In this way the process itself is optimised.

The solution obtained with such a sub-optimal performance function, are often remarkably stable. Figure 13 shows how, for the given example, the optimal D^* and S_{in} vary as a function of the product/substrate price ratio in the performance function. Both D and S_{in} show little change over a broad range. Only close to the point where the performance becomes negative (price ratio = 3.4), their optimal values change. Besides, as the production rate was optimised, the investments cost will be of little influence on the optimal D^* and S_{in}. The performance itself strongly depends on the given price ratio.

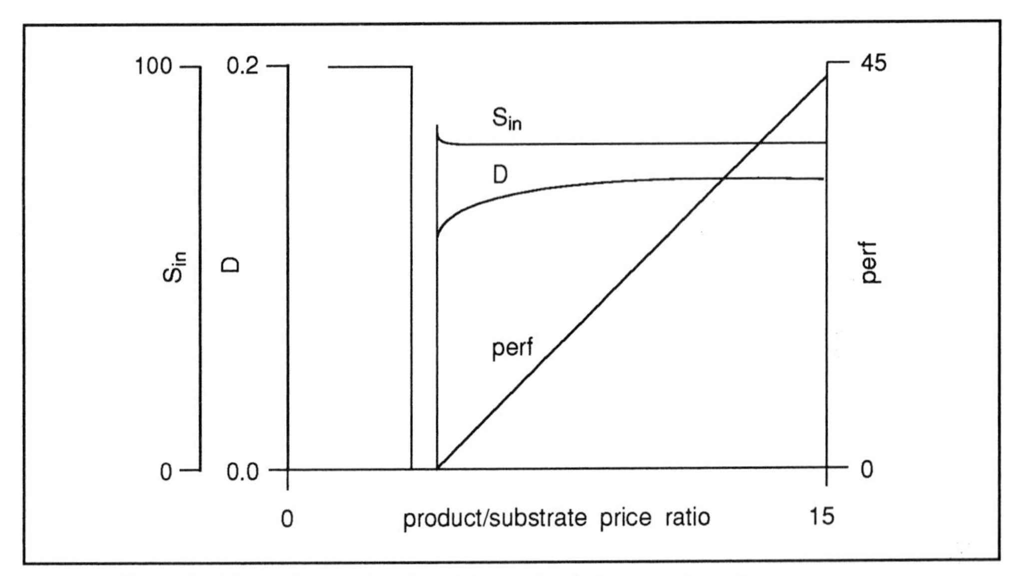

Figure 13 The optimal D and S_{in} as a function of the product/substrate price ratio.

Π Consider the first example. Do you know another, non-iterative method to find the optimal D and S_{in}?

As an algebraic function can be found to calculate the performance as a function of D and S_{in}, one can analytically find the optimum by equating the partial derivatives (to D and S_{in}) of the performance function to zero.

Π Can you figure out why the optimisation of the initial conditions of a batch process will be much more time consuming than the optimisation of the given examples?

For the given examples, each evaluation of the performance function will only cost a single calculation step while the evaluation of the performance function of a batch process involves a complete dynamic simulation of this process, because the performance can only be evaluated at the end of the batch.

Appendix 3

The major symbols used in this text

Typtical units are indicated:

Chapter 2

A = surface area (m^2)

a = specific surface area (m^2 m^{-3})

C = concentration (mass per unit volume)

C_g = concentration in gas phase (mol m^{-3})

C_l = concentration in liquid phase (mol m^{-3})

C_o = initial concentration (mol m^{-3}; kg m^{-3})

C_s = substrate concentration (mol s^{-1})

D = stirrer diameter (m)

D = diffusion coefficient (m^2s^{-1}) [D_e = effective diffusion coefficient]

D_M = stirrer diameter (model) (m)

D_P = stirrer diameter (prototype) (m)

d_p = particle diameter

ε = gas hold up

Fr = Froude number

g = gravitational constant (m s^{-2})

H = height (m)

H_R = stirrer blade height

k_L = mass transfer coefficient (m s^{-1})

m = gas liquid distribution coefficient

N = stirrer speed (rad s^{-1})

OTR = oxygen transfer rate (mol s^{-1})

P = power input (w)

P_g = gassed power input (w)

P_s = ungassed power input (w)

R = gas constant (J mol^{-1} K^{-1})

Re = Reynolds number ($\frac{\rho v D}{\eta}$; $\frac{\rho N D^2}{\eta}$)

r_{hM} = heat production by a micro-organism ($Jm^{-3}s^{-1}$)

r_s = specific substrate consumption rate (mol m^{-3} s^{-1})

T = diameter of a vessel (m)

t_c = circulation time (s)

t_{char} = characteristic time (s)

t_G = characteristic time for growth (s)

t_{hT} = characteristic time for heat transfer (s)

t_m = mixing time (s)

t_{oc} = characteristic time for oxygen consumption (s)

$t_{OT,b}$ = characteristic oxygen transfer time from a bubble (s)

t_{sc} = characteristic time for substrate consumption (s)

U = overall heat transfer coefficient

V = volume (m^3)

V_l = volume of liquid (m^3)

v_{lc} = liquid circulation velocity (m s^{-1})

v_s = gas superficial velocity (m s^{-1})

v_{tip} = stirrer tip velocity (m s^{-1})

α = specific heat transfer coefficient

η = viscosity (Pas)

μ = specific growth rate (s^{-1})

ρ = denisty ($kg\ m^{-3}$)

ϕ_g = gas flow rate ($m^2\ s^{-1}$; $mol\ s^{-1}$)

ϕ_P = pumping capacity ($m^3\ s^{-1}$)

Chapter 3

a = specific surface area ($m^2\ m^{-3}$)

C_s = substrate concentration ($mol\ m^{-3}$)

C_{s0} = initial substrate concentration ($mol\ m^{-3}$)

D = diameter of stirrer (m)

D = dilution rate (s^{-1})

H = height of liquid column (m)

k_L = mass transfer coeficient ($m\ s^{-1}$)

K_s = substrate saturation constant ($mol\ m^{-3}$)

r = specific consumption rate ($mol\ m^{-3}\ s^{-1}$)

r_{O2} = consumption of oxygen

r_s = consumption of substrate

t_G = characteristic time of biomass growth (s)

t_{hP} = characteristic time of heat production (s)

t_{hT} = characteristic time of heat transfer (s)

t_{ml} = characteristic mixing time of the liquid (s)

t_{oc} = characteristic time of oxygen consumption (s)

$t_{OT,l}$ = characteristic oxygen transfer time to the liquid (s)

V = volume (m^3)

τ = residence time (s)

Chapter 4

C_A = mass concentration of species A (kg m^{-3})

C_M = biomass concentration (kg m^{-3})

C_{Mi} = initial biomass concentration (kg m^{-3})

C_{Ml} = biomass concentration in bulk liquid (kg m^{-3})

C_P = concentration of product (kg m^{-3})

C_S = substrate concentration (mol m^{-3}; kg m^{-3})

C_{Si} = input substrate concentration (mol m^{-3}; kg m^{-3})

C_{Sl} = substrate concentration in bulk liquid (mol m^{-3}; kg m^{-3})

C_{So} = output substrate concentration (mol m^{-3}; kg m^{-3})

D = dilution rate (s^{-1})

F = flow rate (m^3 s^{-1})

k_{Pg} = growth associated product formation coefficient (kg kg^{-1})

k_{Pn} = non-growth associated product formation coefficient (kg kg^{-1} s^{-1})

K_s = saturation constant (kg m^{-3}; mol m^{-3})

r = specific rate of reaction (s^{-1}) (subscript M = biomass production, P = product formation)

R_A = rate of reaction of species A per unit volume (kg m^{-3} s^{-1} mol m^{-3} s^{-1})

r_d = specific rate of cell decay (kg kg^{-1} s^{-1})

r_m = specific rate of maintenance energy consumption (kg kg^{-1} s^{-1})

R_{MS} = rate of microbial cell mass production from substrate (kg m^{-3} s^{-1})

R_{PS} = rate of product formation from substrate (kg m^{-3} s^{-1})

R_s = rate of substrate removal (kg m^{-3} s^{-1})

r_s = specific rate of substrate removal (kg kg^{-1} s^{-1})

R_{SM} = rate of substrate mass consumption for production of cell mass (kg m^{-3} s^{-1})

R_{SP} = rate of substrate mass consumption for formation of product mass (kg m^{-3} s^{-1})

V = volume (m^3)

Y_{MSg} = true growth yield (kg kg^{-1})

Y_{MSo} = observed microbial cell yield or observed growth yield (kg kg^{-1})

Y_{PS} = product yield (kg kg^{-1})

μ = specific growth rate (s^{-1})

Chapter 5

C = concentration (kg m^{-3}; mol m^{-3}), subscript M = biomass, P = product, S = substrate, i = input (initial), l = bulk liquid

F = flow rate (m^3 s^{-1})

t = time (s)

V = volume (m^3)

r = specific rate of reaction (kg kg^{-1} s^{-1}), subscript M = biomass, p = product, s = substrate

Chapter 6

A = area (m^2)

C = concentration (kg m^{-3}; mol m^{-3}), subscript M = biomass, P = product, s = substrate, i = input

k_{pg} = growth associated product yield (kg kg^{-1})

k_{pn} = non growth associated product yield (kg kg s^{-1})

K_s = substrate saturation constant (kg m^{-3}; mol m^{-3})

R = rate of consumption/formation (kg m^{-3} s^{-1}; mol m^{-3} s^{-1}), subscripts M = biomass, P = product, S = substrate

r = specific rate of production formation (kg m^{-3} s^{-1})

v = velocity (m s^{-1})

Y_{MSo} = observed biomass yield (kg kg^{-1})

Y_{PS} = product yield (kg kg^{-1})

μ = specific growth rate (s^{-1})

Chapter 7

C = concentration (kg m^{-3}), subscript M = biomass, S = substrate, P = volume, i = input, l = bulk liquid

D = dilution rate (s^{-1})

F = flow rate (m^{3} s^{-1})

F_c = flow rate from a concentrator (m^{3} s^{-1})

r_P = specific rate of product formation (kg m^{-3} s^{-1})

V = volume (m^{3})

V_c = concentrator volume

V_R = reactor volume (m^{3})

Y_{MSo} = observed biomass yield (kg kg^{-1})

Y_{ps} = product yield (kg kg^{-1})

α = recycled fraction of F

β = concentration factor

θ_b = mean cell residence time or sludge age (s)

θ_h = hydraulic residence time (s)

μ = specific growth rate (s^{-1})

Chapter 8

C = concentration (kg m^{-3}), subscript M = biomass, P = product, S = substrate, f = biofilm, l = bulk liquid

D = dilution rate (s^{-1})

D_s = mass diffusion coefficient of substrate (m^{2} s^{-1}), subscript f = biofilm, l = liquid

k_{Pg} = growth associated product yield (coefficient) (kg kg^{-1})

k_{Pn} = non-growth associated product yield (coefficient) (kg kg^{-1} s^{-1})

r = specific rate (kg kg^{-1} s^{-1}), subscript Md = biomass detachment, Mf = biomass growth in a biofilm, P = product, S = substrate

V = volume (m^3)

Y_{MSo} = observed growth yield ($kg\ kg^{-1}$)

Y_{PS} = product yield ($kg\ kg^{-1}$)

μ = specific growth rate (s^{-1}), subscript f = biofilm, l = bulk liquid

ϕ = flux ($kg\ m^{-2}\ s^{-1}$), subscript M = biomass, P = product, S = substrate, f = biofilm, l = bulk liquid

Chapter 9

r = conversion rate ($kg\ kg^{-1}\ s^{-1}$), subscripts c = carbon dioxide, n = nitrogen, o = oxygen, P = product, x = biomass

Appendix 4

Recommended values of physical constants

Physical constant	Symbol	Value
acceleration due to gravity	g	$9.81\ m\ s^{-2}$
Avogadro constant	N_A	$6.022.10^{23}\ mol^{-1}$
Boltzman constant	k	$1.380\ 10^{23}\ JK^{-1}$
Faraday constant	F	$9.649.10^{4}\ C\ mol^{-1}$
gas constant	R	$8.314\ JK^{-1}\ mol^{-1}$
molar volume of ideal gas of stp	V_m	$2.241.10^{-2}\ m^3\ mol^{-1}$
Planck constant	h	$6.628.10^{-34}\ Js$
velocity of light in a vacuum	c	$2.998\ 10^{8}\ m\ s^{-1}$

Appendix 5 - Dimensional Numbers

This appendix contains some frequently used dimensionless numbers, their definition and their meaning.

symbol	name	formula	meaning	used in
Ah	Arrhenius	$\frac{E}{RT}$	$\frac{\text{activation energy}}{\text{thermal energy}}$	chemical reactions
Bo	Bodenstein	$\frac{vL}{D}$	$\frac{\text{convection}}{\text{axial diffusion}}$	Bo = Pe diffusion in reactors
Bd	Bond	$\frac{\Delta\rho D^2 g}{\sigma}$	$\frac{\text{gravitational force}}{\text{surface tension}}$	Bd = Eo bubbles and drops
Da I	Damkohler I	$\frac{k_r D}{v}$	$\frac{\text{chemical reaction rate}}{\text{convection mass transfer rate}}$	chemical reactions
Da II	Damkohler II	$\frac{k_r D^2}{\mathcal{D}}$	$\frac{\text{chemical reaction rate}}{\text{diffusive mass transfer rate}}$	chemical reactions
Eo	Eotvos	$\frac{\Delta\rho D^2 g}{\sigma}$	$\frac{\text{gravitational force}}{\text{surface tension}}$	Eo = Bd bubbles and drops
f	Fanning fricition factor	$\frac{d\Delta p}{2\rho v^2 L}$	$\frac{\text{shear stress energy at the wall}}{\text{kinetic energy}}$	flow through tubes and channels
Fo	Fourier (mass)	$\frac{\mathcal{D}t}{d^2}$	$\frac{\text{process time}}{\text{diffusion time}}$	diffusion
Fo	Fourier (heat)	$\frac{\alpha t}{D^2}$	$\frac{\text{process time}}{\text{conduction time}}$	heat conduction
Fr	Froude	$\frac{v^2}{gH}$	$\frac{\text{inertia forces}}{\text{gravity forces}}$	
Fr	Froude (rotation)	$\frac{DN^2}{g}$	$\frac{\text{inertia forces}}{\text{gravity forces}}$	mixing with free surfaces
Gz	Graetz	$\frac{\alpha L}{D^2 v}$	$\frac{\text{conductive heat transfer}}{\text{convective heat transfer}}$	heat transfer to flowing media
Gr	Grashof	$\frac{D^3 g}{v^2}\frac{\Delta\rho}{\rho}$	$\frac{\text{buoyancy forces}}{\text{viscous forces}}$	free convection
Ha	Hatta	$\frac{\sqrt{\mathcal{D} k_r}}{k}$	$\frac{\text{mass transfer with chemical reaction}}{\text{mass transfer without chemical reaction}}$	mass transfer with chemical reaction

Le	Lewis	$\frac{\alpha}{D}$	$\frac{\text{thermal boundary layer thickness}}{\text{mass transfer boundary layer thickness}}$	$Pe = \frac{Sc}{Pr}$ combined heat and mass transfer
Nu	Nusselt	$\frac{hD}{\lambda}$	$\frac{\text{total heat transfer}}{\text{conductive heat transfer}}$	heat transfer
Pe	Peclet (heat)	$\frac{vD}{\alpha}$	$\frac{\text{convective heat transfer}}{\text{conductive heat transfer}}$	heat transfer in flowing media
Pe	Peclet (mass)	$\frac{vd}{D}$	$\frac{\text{convective mass transfer}}{\text{diffusive mass transfer}}$	mass transfer in flowing media
Po	Power number	$\frac{P}{\rho N^3 D^5}$	$\frac{\text{power added}}{\text{power transferred to kinetic energy}}$	stirred vessels and pumps
Pr	Prandtl	$\frac{v}{\alpha}$	$\frac{\text{hydrodynamic boundary layer thickness}}{\text{thermal boundary layer thickness}}$	heat transfer in flowing media
Re	Reynolds	$\frac{\rho vD}{\eta}$	$\frac{\text{inertia forces}}{\text{viscous forces}}$	flow
Sc	Schmidt	$\frac{v}{D}$	$\frac{\text{hydrodynamic boundary layer thickness}}{\text{mass transfer boundary layer thickness}}$	mass transfer in flowing media
Sh	Sherwood	$\frac{kD}{D}$	$\frac{\text{total mass transfer}}{\text{diffusive mass transfer}}$	mass transfer
T	Thiele modulus	$D\sqrt{\frac{k_r}{D}}$	$\frac{\text{chemical reaction rate}}{\text{diffusive mass transfer}}$	$T = \sqrt{DaII}$ chemical reactions
We	Weber	$\frac{\rho v^2 d}{\sigma}$	$\frac{\text{inertia forces}}{\text{surface tension forces}}$	We = Eo . Fr bubbles and drops

Index

A

B

C

T

U

V

W

X

Y

Z